Lecture Notes in Statistics 192

Edited by P. Bickel, S. Fienberg

David B. Dunson
Editor

Random Effect and Latent Variable Model Selection

Springer

Editor
David B. Dunson
National Institute of Environmental Health Sciences
Research Triangle Park, NC
USA
dunson@stat.duke.edu

ISBN: 978-0-387-76720-8 e-ISBN: 978-0-387-76721-5
DOI: 10.1007/978-0-387-76721-5

Library of Congress Control Number: 2008928920

Cover illustration: Follicles of colloid in thyroid

Printed on acid-free paper

9 8 7 6 5 4 3 2 1

springer.com

Preface
Random Effect and Latent Variable Model Selection

In recent years, there has been a dramatic increase in the collection of multivariate and correlated data in a wide variety of fields. For example, it is now standard practice to routinely collect many response variables on each individual in a study. The different variables may correspond to repeated measurements over time, to a battery of surrogates for one or more latent traits, or to multiple types of outcomes having an unknown dependence structure. Hierarchical models that incorporate subject-specific parameters are one of the most widely-used tools for analyzing multivariate and correlated data. Such subject-specific parameters are commonly referred to as random effects, latent variables or frailties.

There are two modeling frameworks that have been particularly widely used as hierarchical generalizations of linear regression models. The first is the linear mixed effects model (Laird and Ware , 1982) and the second is the structural equation model (Bollen , 1989). Linear mixed effects (LME) models extend linear regression to incorporate two components, with the first corresponding to fixed effects describing the impact of predictors on the mean and the second to random effects characterizing the impact on the covariance. LMEs have also been increasingly used for function estimation. In implementing LME analyses, model selection problems are unavoidable. For example, there may be interest in comparing models with and without a predictor in the fixed and/or random effects component. In addition, there is typically uncertainty in the subset of predictors to be included in the model, with the number of candidate predictors large in many applications.

To address problems of this type, it is not appropriate to rely on classical methods developed for model selection and inferences in non-hierarchical regression models. For example, the widely used BIC criteria are not valid for random effects models, and likelihood ratio and score tests face difficulties, since the null hypothesis often falls on the boundary of the parameter space. The objective of the first part of this book is to provide an overview of a variety of promising strategies for addressing model selection problems in LMEs and related modeling frameworks.

In the chapter, "Likelihood Ratio Testing for Zero Variance Components in Linear Mixed Models," Ciprian Crainiceanu provides an applications-motivated overview of recent work on likelihood ratio and restricted likelihood ratio tests for testing whether random effects have zero variance. The approaches he describes represent an important advance over the current standard practice in testing for zero variance components in hierarchical models. Such approaches include ignoring the boundary problem and assuming the likelihood ratio test statistic has a chi-square distribution under the null and relying on asymptotic results showing a mixture of chi-squares is more appropriate (Stram and Lee, 1994). Crainiceanu shows that asymptotic approximations may be unreliable in many applications, motivating use of finite sample approaches. He illustrates the ideas through several examples, including applications to nonlinear regression modeling.

Score tests provide a widely-used alternative to likelihood ratio tests, and in the chapter, "Variance Component Testing in Generalized Linear Mixed Models for Longitudinal/Clustered Data and Other Related Topics," of this volume Daowen Zhang and Xihong Lin provide an excellent overview of the recent literature on score test-based approaches. In addition, Zhang and Lin consider a broader class of models, which includes GLMMs and generalized additive mixed models (GAMMs). GAMMs provide an extremely rich framework for semiparametric modeling of longitudinal data allowing flexible predictor effects through replacing linear terms in a generalized linear model with unknown non-linear functions, while also including random effects to account for within-subject dependence and heterogeneity.

The first part of the volume is completed with two companion chapters describing Bayesian approaches for variable selection in LMEs and GLMMs. The likelihood ratio and score test methods provide an approach for comparing two nested models with the smaller model having a random effect excluded. However, in many applications one is faced with a set of p candidate predictors, with uncertainty in which subsets should be included in the fixed and random effects components of the model. Clearly, the number of candidate models grows extremely rapidly with p, so that it often becomes impossible to fit each model in the list. One possibility is to use a likelihood ratio test within a stepwise selection procedure. However, the final model selected will depend on the order in which candidate predictors are added or deleted and it is difficult to adjust for uncertainty in subset selection in performing inferences and predictions. In non-hierarchical regression models, Bayesian variable selection implemented with stochastic search algorithms has been very widely used to address this problem. In the chapter, "Bayesian Model Uncertainty in Mixed Effects Models," Satkartar Kinney and I describe an approach for LMEs, while in the chapter, "Bayesian Variable Selection in Generalized Linear Mixed Models," Bo Cai and I describe an alternative for GLMMs.

The second part of the book switches gears to focus on structural equation models (SEMs), which have been very widely used in social science applications for assessing relationships among latent variables, such as poverty or violence, that can only be measured indirectly through multiple surrogates. SEMs provide a generalization of factor analysis, which allows for modeling of linear relationships among the latent factors through a linear structural relations (LISREL) model. SEMs are also quite useful outside of traditional application areas for sparse covariance structure modeling of high-dimensional multivariate data. However, one of the main issues in applying SEMs is how to deal with model uncertainty, which commonly arises in deciding on the number of factors to include in each component and the relationships among these factors. In the chapter, "A Unified Approach to Two-Level Structural Equation Models and Linear Mixed Effects Models," Peter Bentler and Jiajuan Liang provide a bridge between the first and second parts of the volume in linking LMEs and SEMs, while also considering methods for model selection.

In the chapter, "Bayesian Model Comparison of Structural Equation Models," Sik-Yum Lee and Xin-Yuan Song provide a general Bayesian approach to comparison of SEMs. Typical Bayesian methods for comparing models rely on Bayes factors. However, Bayes factors have proved quite difficult to estimate accurately in SEMs. Lee and Song propose a useful and clever solution to this problem using path sampling. One well-known issue in model selection using Bayes factors is sensitivity to prior selection. This has motivated a rich literature on default priors. In the chapter, "Bayesian Model Selection in Factor Analytic Models" Joyee Ghosh and I build on the approach of Lee and Song, proposing a default prior, and an efficient approach for posterior computation relying on parameter expansion. In addition, an importance sampling algorithm is proposed as an alternative to path sampling.

In summary, this volume provides a practically-motivated overview of a variety of recently proposed approaches for model selection in random effects and latent variable models. The goal is to make these methods more accessible to practitioners, while also stimulating additional research in this important and under-studied area of statistics. There are a number of topics related to model selection in random effects and latent variable models that are in need of new research, with solutions having the potential for substantial applied impact. The first topic is the development of simple methods to calculate model selection criteria, which modify AIC and BIC to incorporate a penalty for model complexity that is appropriate for a hierarchical model. A second topic is the development of efficient methods for simultaneous model search and posterior computation in SEMs. Often, one has a high-dimensional set of SEMs that are plausible a priori and consistent with current scientific or sociologic theories. It is of substantial interest to identify high posterior probability models and to average across models in making predictions. However, typical tricks used in other model classes, such as zeroing out coefficients, do not work in general for SEMs, and efficient alternatives remain to be developed.

References

Bollen, K.A. (1989). *Structural Equation Models with Latent Variables*. New York: Wiley

Laird, N. and Ware, J. (1982). Random-effects models for longitudinal data. *Biometrics* **38**, 963–974

David B. Dunson

Contents

Part I
Random Effects Models

Likelihood Ratio Testing for Zero Variance Components in Linear Mixed Models

Ciprian M. Crainiceanu

Mixed models are a powerful inferential tool with a wide range of applications including longitudinal studies, hierarchical modeling, and smoothing. Mixed models have become the state of the art for statistical information exchange and correlation modeling. Their popularity has been augmented by the availability of dedicated software, e.g., the `MIXED` procedure in SAS, the `lme` function in R and S+, or the `xtmixed` function in STATA.

In this paper, we consider the problem of testing the null hypothesis of a zero variance component in a linear mixed model (LMM). We focus on the likelihood ratio test (LRT) and restricted likelihood ratio test (RLRT) statistics for three reasons. First, (R)LRTs are uniformly most powerful for simple null and alternative hypotheses and have been shown to have good power properties in a variety of theoretical and applied frameworks. Second, given their robust properties, (R)LRTs are the benchmark for statistical testing. Third, (R)LRT can now be used in realistic data sets and applications due to a better understanding of their null distribution and improved computational tools.

The paper is organized as follows. Section 1 describes three applications of testing for a zero variance component. Section 2 contains the model and a description of the testing framework. Section 3 describes standard asymptotic results and provides a short discussion of their applicability. Section 4 presents finite sample and asymptotic results for linear mixed models (LMMs) with one variance component. Section 5 introduces two approximations of the finite sample (R)LRT distribution for testing for zero variance components in LMMs with multiple variance components. Section 6 presents the corresponding testing results for the examples introduced in Sect. 1. Section 7 provides the discussion and practical recommendations.

C.M. Crainiceanu
Department of Biostatistics, Johns Hopkins University
ccrainic@jhsph.edu

D. B. Dunson (ed.) *Random Effect and Latent Variable Model Selection*,
DOI: 10.1007/978-0-387-76721-5,

1 Examples

The three examples in this section illustrate the wide variety of applications of testing for zero variance components in LMMs. This list is far from being exhaustive but provides a foretaste of what is possible and needed in this framework.

1.1 Loa loa *Prevalence in West Africa*

Figure 1 displays village locations from one of the several parasitological survey location in West Africa. In all these villages parasitological sampling was conducted to assess the prevalence of Loaisis. Here we provide a short summary, but a complete description of the problem can be found in Crainiceanu et al. (2007). Loaisis, or eyeworm, is an endemic disease of the wet tropics, caused by *Loa loa*, a filarial parasite which is transmitted to humans by the bite of an infected *Chrysops* fly. In Fig. 1 the empirical prevalence rates at location $\boldsymbol{x}$, $\widehat{p}(\boldsymbol{x})$, are indicated as dots coded according to their size: small $\widehat{p}(\boldsymbol{x}) < 0.18$, medium $0.18 \leq \widehat{p}(\boldsymbol{x}) < 0.20$, large $0.20 \leq \widehat{p}(\boldsymbol{x}) < 0.25$, and very large $\widehat{p}(\boldsymbol{x}) > 0.30$.

A complete bivariate binomial analysis of this data set can be found in Crainiceanu et al. (2007). Here, we consider the following simpler univariate model for the logit prevalence at the spatial location $\boldsymbol{x}$

$$\text{logit}\{\widehat{p}(\boldsymbol{x})\} = \alpha_0+\alpha_1 g(\boldsymbol{x})+\alpha_2 s(\boldsymbol{x})+\alpha_3 e(\boldsymbol{x})+\alpha_4\{e(\boldsymbol{x})-800\}_+ +S(\boldsymbol{x})+\epsilon(\boldsymbol{x}), \quad (1)$$

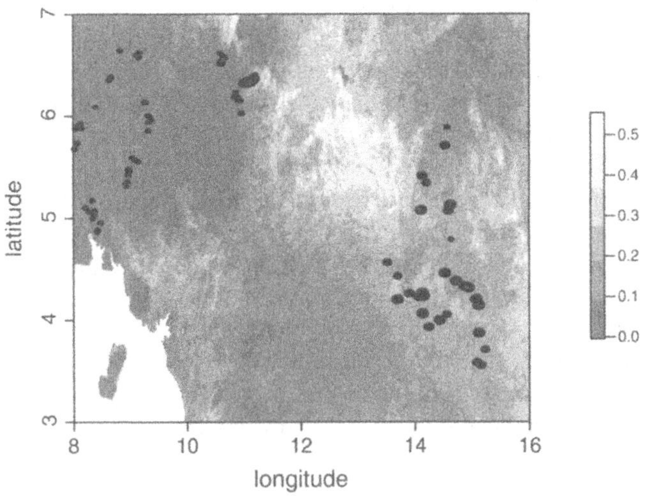

Fig. 1 Village sampling locations in one subregion from West Africa. The empirical prevalence rates are indicated as *dots* coded according to their size: small $\widehat{p}(\boldsymbol{x}) < 0.18$, medium $0.18 \leq \widehat{p}(\boldsymbol{x}) < 0.20$, large $.20 \leq \widehat{p}(\boldsymbol{x}) < 0.25$, very large $\widehat{p}(\boldsymbol{x}) > 0.30$. The estimated mean prevalence based on model (1) is grey-scale coded according to the legend

where $g(\boldsymbol{x})$ is an annual average measure of greenness, $s(\boldsymbol{x})$ the standard deviation of greenness, $e(\boldsymbol{x})$ the elevation in meters, $S(\boldsymbol{x})$ a spatial component, and $\epsilon(\boldsymbol{x}) \sim \text{Normal}(0, \sigma_\epsilon^2)$ are the independent errors. Here a_+ is equal to a if $a > 0$ and 0 otherwise, so that $\{e(\boldsymbol{x}) - 800\}_+$ represents the elevation at location $\boldsymbol{x}$ truncated below 800 m. If the spatial component $S(\boldsymbol{x})$ is modeled as a low rank penalized thin plate spline then

$$\begin{cases} S(\boldsymbol{x}) = \boldsymbol{x}^t\boldsymbol{\beta} + \boldsymbol{Z}(\boldsymbol{x})\boldsymbol{b}, \\ \boldsymbol{b} \quad \sim \text{Normal}(0, \sigma_b^2 I_K), \end{cases} \tag{2}$$

where $Z(\boldsymbol{x})$ is the low rank specific design vector (for details see (Ruppert et al., 2003; Kammann and Wand, 2003)), b the thin plate spline coefficients describing the spatial features of $S(\boldsymbol{x})$, σ_b^2 the smoothing parameter controlling the amount of smoothing, and I_K is the identity matrix where K is the number of spatial knots.

In the case of low rank smoothers the set of K knots for the covariates have to be chosen. One possibility is to use equally spaced knots. Another possibility is to select the knots and subknots using the *space filling* design (Nychka and Saltzman, 1998), which is based on the maximal separation principle. This avoids wasting knots and is likely to lead to better approximations in sparse regions of the data. The `cover.design()` function from the R package `Fields` (Fields Development Team, 2006) provides software for space filling knot selection.

If the smoothing parameter is estimated by restricted maximum likelihood (REML), then the model described in (1) and (2) is equivalent to a particular LMM with one variance component. Figure 1 displays the estimated mean prevalence at all locations in the map coded according to the legend. In this context testing whether the nonlinear spatial component of $S(\boldsymbol{x})$ is necessary to explain the residual variability after fitting the scientifically available covariates is equivalent to testing

$$H_0 : \sigma_b^2 = 0 \quad \text{vs.} \quad H_A : \sigma_b^2 > 0 \,.$$

From a scientific perspective testing H_0 is equivalent to testing whether simpler models including only covariates could capture the complex stochastic nature of the spatial data and have good predictive power.

1.2 Onion Density in Australia

Figure 2 contains data on yields (grams/plant) of white Spanish onions in two locations: Purnong Landing and Virginia, South Australia (Ratkowsky, 1983). The horizontal axis corresponds to areal density of plants (plants/m^2). Detailed analyses of these data are given by Ruppert et al. (2003) and Crainiceanu (2003). Denote by (y_i, x_i, s_i) the yield, density of plants and location for the ith observation. Here, $s_i = 1$ corresponds to Purnong Landing and $s_i = 0$ corresponds to Virginia. The solid lines in Fig. 2 correspond to fitting the linear additive model

$$\log(y_i) = \beta_0 + \beta_1 s_i + \beta_2 d_i + \epsilon_i \,. \tag{3}$$

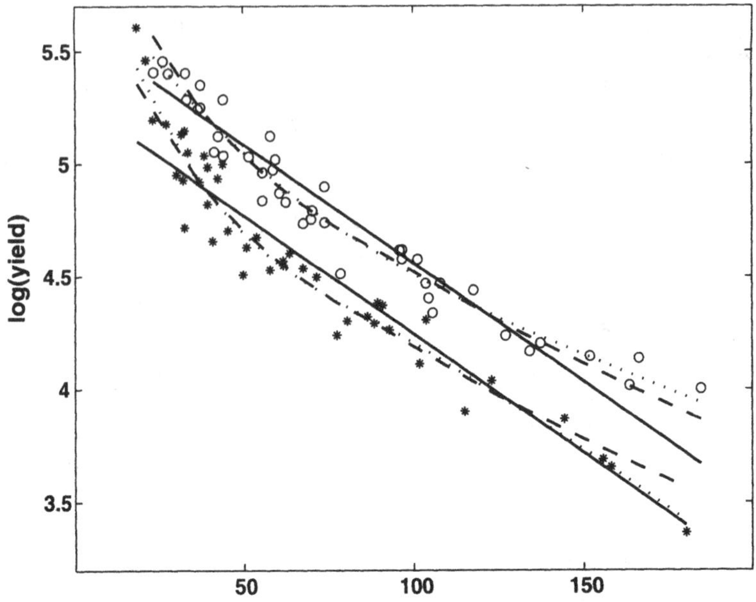

Fig. 2 Log yield for the onion data plotted against density (*circle* Purnong Landing; *asterisk* Virginia), straight line fit (*solid line*), binary offset model using a penalized linear spline fit with $K = 15$ knots and REML estimation of smoothing parameter (*dashed line*), discrete by continuous interaction model (*dotted line*)

The dashed lines represent the mean fit using a *semiparametric binary offset model* (Ruppert et al., 2003)

$$\log(y_i) = \beta_1 s_i + f(d_i) + \epsilon_i, \tag{4}$$

which contains a parametric component, $\beta_1 s_i$, and a nonparametric component, $f(d_i)$. The binary variable s vertically offsets the relationship between $E[\log(y)]$ and density according to location. By specifying a linear penalized spline model for $f(d_i)$ the model becomes

$$\log(\mathrm{y}_i) = \beta_0 + \beta_1 s_i + \beta_2 d_i + \sum_{k=1}^{K} b_k (d_i - \kappa_k)_+ + \epsilon_i,$$

where b_k are i.i.d. $N(0, \sigma_b^2)$ and ϵ_i are i.i.d. $N(0, \sigma_\epsilon^2)$. Following Ruppert et al. (2003), we use $K = 15$ knots chosen at the sample quantiles of density corresponding to frequencies $1/(K+1), \ldots, K/(K+1)$.

Testing model (3) corresponding to the solid line fits in Fig. 2 versus model (4) corresponding to dashed lines in Fig. 2 corresponds to testing $H_0 : \sigma_b^2 = 0$ vs. $H_A : \sigma_b^2 > 0$. For these data and hypothesis testing framework, Crainiceanu (2003) calculated RLRT $= 35.93$ with a corresponding p-value < 0.001. The calculation of the p-value was based on the exact distribution of the RLRT as obtained by Crainiceanu and Ruppert (2004b). This result is not surprising, given the large discrepancies between the two model fits in Fig. 2. In fact, results would not change even if one used the more conservative (but incorrect in this case) $0.5\chi_0^2 : 0.5\chi_1^2$ approximation to the null RLRT distribution (Self and Liang, 1987).

It is natural, however, to ask whether the binary offset model accurately represents the data. To address this question we nest model (4) into the following discrete by continuous interaction model

$$E\{\log(y_i)\} = \begin{cases} f_{\mathrm{PL}}(d_i) & \text{if } s_i = 1; \\ f_{\mathrm{VA}}(d_i) & \text{if } s_i = 0, \end{cases}$$

where the subscripts PL and VA denote the Purnong Landing and Virginia locations, respectively. The basic idea is to model the mean response at one of the locations, say Purnong Landing, as a nonparametric spline and the deviations from this function corresponding to the other location, say Virginia, as another nonparametric spline. The discrete by continuous interaction model is

$$\log(y_i) = \beta_0 + \beta_1 d_i + \sum_{k=1}^{K} b_k(d_i - \kappa_k)_+ + \{\gamma_0 + \gamma_1 d_i + \sum_{k=1}^{K} v_k(d_i - \kappa_k)_+\} I(i \in \mathrm{PL}) + \epsilon_i \tag{5}$$

for Virginia ($s = 0$), where β_0, β_1, γ_0, and γ_1 are fixed unknown parameters, b_k are i.i.d. $N(0, \sigma_b^2)$, v_k are i.i.d. $N(0, \sigma_v^2)$, and $I(i \in \mathrm{PL})$ is 1 if the observation i is from Purnong Landing and 0 otherwise. The model (5) is an LMM with two random effects variance components, σ_b^2 and σ_v^2, and the fit to the data is depicted by the two dotted curves in Fig. 2. Testing for linear versus nonlinear deviations from the smooth regression function corresponding to the Purnong Landing location reduces in this model to testing

$$H_0 : \sigma_v^2 = 0 \quad \text{vs.} \quad \sigma_v^2 > 0,$$

which is equivalent to testing for a zero variance component in an LMM with two variance components. After discussing the state of the art in statistical testing in this framework we will revisit this example in Sect. 6.

1.3 Coronary Sinus Potassium

We consider the coronary sinus potassium concentration data measured on 36 dogs published by Grizzle and Allan (1969) and Wang (1998). The measurements on each dog were taken every 2 min from 1 to 13 min (seven observations per dog). The 36 dogs come from four treatment groups. Figure 3 displays the data for the nine dogs in the first treatment group (dotted lines).

If y_{ij} denotes the jth concentration for the ith dog at time $t_{ij} = 1 + 2j$ then a reasonable LMM model for the first treatment group is

$$y_{ij} = \beta_0 + u_i + \beta_1 t_{ij} + \beta_2 t_{ij}^2 + \epsilon_{ij}, \tag{6}$$

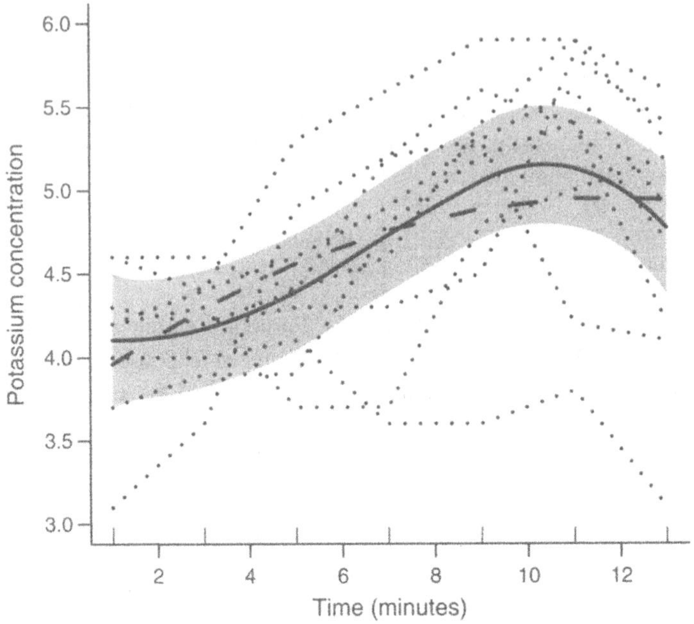

Fig. 3 Sinus potassium concentration for nine dogs in the first treatment group (*dotted lines*)

where $u_i \sim N(0, \sigma_u^2)$ are independent dog specific intercepts and $\epsilon_{ij} \sim N(0, \sigma_\epsilon^2)$ are independent errors. Figure 3 displays the fit of model (6) as a dashed line. It is natural to ask the question whether model (6) is enough to capture the complexity of the population mean function. One way to answer this question is by embedding model (6) into the following more general model

$$y_{ij} = \beta_0 + u_i + \beta_1 t_{ij} + \beta_2 t_{ij}^2 + \sum_{i=1}^{K} b_k (t_{ij} - \kappa_k)_+^2 + \epsilon_{ij}, \tag{7}$$

where $b_k \sim N(0, \sigma_b^2)$ are independent truncated spline coefficients, K the number of knots and κ_k, $k = 1, \ldots, K$ are the knots. All the other assumptions are the same as in model (6). Note that model (6) is an LMM with two variance components: one, σ_u^2, controlling the shrinkage of random intercepts towards their mean and the other one, σ_b^2, controlling the shrinkage of the population function towards a quadratic polynomial. Figure 3 displays the fit of this model as a solid line together with 95% pointwise confidence intervals (shaded area).

Testing the null hypothesis described by model (6) versus the alternative described by model (7) is equivalent to testing for

$$H_0 : \sigma_b^2 = 0 \quad \text{vs.} \quad \sigma_b^2 > 0.$$

Similarly, testing for dog response homogeneity is equivalent to testing

$$H_0 : \sigma_u^2 = 0 \quad \text{vs.} \quad \sigma_u^2 > 0.$$

Both frameworks correspond to testing for a zero variance component in an LMM with two variance components.

As the last point for this example, note that a naive way to test for $H_0 : \sigma_b^2 = 0$ is to check whether the null fit is contained in the shaded area. This may seem like a good idea, but leads to incorrect inferences. Indeed, all the confidence intervals for the mean function based on model (7) contain the fit based on model (6). However, as we show in Sect. 6, the RLRT indicates strong evidence against the null hypothesis of a quadratic population curve.

2 Model and Testing Framework

All examples in Sect. 1, and many others, involve testing for a zero variance component as the methodological answer to important scientific questions. To formalize the framework, let us assume that the outcome vector, $\boldsymbol{Y}$, is modeled as an LMM

$$\begin{cases} \boldsymbol{Y} = \boldsymbol{X\beta} + \boldsymbol{Z}_1\boldsymbol{b}_1 + \cdots + \boldsymbol{Z}_S\boldsymbol{b}_S + \boldsymbol{\epsilon}, \\ \boldsymbol{b}_s \sim N(\boldsymbol{0}, \sigma_s^2 \boldsymbol{I}_{K_s}), \quad s = 1, \ldots, S, \\ \boldsymbol{\epsilon} \sim N(\boldsymbol{0}, \sigma_\epsilon^2 \boldsymbol{I}_n). \end{cases} \tag{8}$$

Here the random effects $\boldsymbol{b}_s$, $s = 1, \ldots, S$, and the error vector $\boldsymbol{\epsilon}$ are mutually independent, K_s denotes the number of columns in $\boldsymbol{Z}_s$, n the sample size, and $\boldsymbol{I}_\nu$ denotes the identity matrix with ν columns. This is not the most general form of an LMM, but it is often used in practice and keeps the presentation simple.

We are interested in testing

$$H_{0,s} : \sigma_s^2 = 0 \quad \text{vs.} \quad H_{A,s} : \sigma_s^2 > 0, \tag{9}$$

where the hypotheses are indexed by $s = 1, \ldots, S$ to emphasize that these are distinct and not joint hypotheses for all variance components. Note that because $\boldsymbol{b}_s \sim N(\boldsymbol{0}, \sigma_s^2 \boldsymbol{I}_{K_s})$, the null hypothesis is equivalent to $\boldsymbol{b}_s = \boldsymbol{0}$, indicating that under the null the component $\boldsymbol{Z}_s\boldsymbol{b}_s$ of model (8) is zero.

Denote by $\boldsymbol{\theta}_{-s}$ all the parameters in model (8) with the exception of σ_s^2. The RLRT for testing $H_{0,s}$ is then defined as

$$\text{RLRT} = 2\text{sup}_{\theta_{-s},\sigma_s^2}\{\log L(\theta_{-s}, \sigma_s^2)\} - 2\text{sup}_{\theta_{-s}}\{\log L(\theta_{-s}, 0)\},$$

where $L(\theta_{-s}, \sigma_s^2)$ is the restricted likelihood function for model (8). A similar definition holds for LRT using the likelihood instead of the restricted likelihood function.

3 Standard Asymptotic Results for LMMs

Testing for zero variance components is not new in mixed models. Using theory originally developed by Chernoff (1954), Moran (1971), and Self and Liang (1987), Stram and Lee (1994) proved that the LRT for testing (9) has an asymptotic $0.5\chi_0^2 : 0.5\chi_1^2$ mixture distribution under the null hypothesis $H_{0,s}$ if data are independent and identically distributed *both under the null and alternative hypothesis.* For more details on standard asymptotic results, see the chapter by Zhang and Lin (2007) in this book. Thus, it could be surprising that in many applications the null distribution of the LRT using simulations is far from being a $0.5\chi_0^2 : 0.5\chi_1^2$ mixture.

There are several reasons for these inconsistencies. First, the Laird and Ware (1982) model used by Stram and Lee (1994) allows the partition of the outcome vector $\boldsymbol{Y}$ into independent subvectors. This could be revealed by close inspection of this model, which is typically described in terms of the subject-level vector $\boldsymbol{Y}_i$ and not in terms of the data vector $\boldsymbol{Y}$. The independence assumption is violated, for example, when representing nonparametric smoothing as a particular LMM. Second, even when the outcome vector can be partitioned into independent subvectors, the number of subvectors may not be sufficient to ensure an accurate asymptotic approximation. Third, subvectors may not be identically distributed due to unbalanced designs or missing data. In the case of an LMM with one variance component ($S = 1$) Crainiceanu and Ruppert (2004b) and Crainiceanu et al. (2005) have derived the finite sample and asymptotic distribution of the LRTs showing that, under general conditions, the null distribution for testing $H_{0,s}$ is typically different from $0.5\chi_0^2 : 0.5\chi_1^2$. In the following section, we provide a summary of these results and discuss the implications for applied statistical inference.

4 Finite Sample and Asymptotic Results for General Design LMMs with One Variance Component

Consider the particular case of model (8) with Gaussian outcome vector and one variance component

$$\begin{cases} \boldsymbol{Y} &= \boldsymbol{X}\boldsymbol{\beta} + \boldsymbol{Z}_1\boldsymbol{b}_1 + \boldsymbol{\varepsilon}, \\ \boldsymbol{b}_1 &\sim N(\boldsymbol{0}, \sigma_1^2 \boldsymbol{I}_{K_1}), \\ \boldsymbol{\varepsilon} &\sim N(\boldsymbol{0}, \sigma_\varepsilon^2 \boldsymbol{I}_n), \end{cases} \tag{10}$$

where $\boldsymbol{b}_1$ and $\boldsymbol{\varepsilon}$ as mutually independent.

As model (10) has only one variance component, σ_1^2, the exact null distribution of the RLRT for testing $H_{0,1} : \sigma_1^2 = 0$ versus $H_{A,1} : \sigma_1^2 > 0$ is Crainiceanu and Ruppert (2004b)

$$\mathrm{RLRT}_n \stackrel{d}{=} \sup_{\lambda \geq 0} \left\{ (n-p)\log\left[1 + \frac{N_n(\lambda)}{D_n(\lambda)}\right] - \sum_{l=1}^{K_1} \log(1 + \lambda\mu_{l,n}) \right\}, \tag{11}$$

where "$\overset{d}{=}$" denotes equality in distribution, p is the number of columns in X,

$$N_n(\lambda) = \sum_{l=1}^{K_1} \frac{\lambda \mu_{l,n}}{1 + \lambda \mu_{l,n}} w_l^2, \quad D_n(\lambda) = \sum_{l=1}^{K_1} \frac{w_l^2}{1 + \lambda \mu_{l,n}} + \sum_{l=K_1+1}^{n-p} w_l^2,$$

$w_l, l = 1, \ldots, n - p$, are independent $N(0, 1)$, and $\mu_{l,n}, l = 1, \ldots, K_1$, are the eigenvalues of the $K_1 \times K_1$ matrix $\boldsymbol{Z_1}'(\boldsymbol{I_n} - \boldsymbol{X}(\boldsymbol{X}'\boldsymbol{X})^{-1}\boldsymbol{X}')\boldsymbol{Z_1}$. The asymptotic distribution of the LRT was also derived by Crainiceanu and Ruppert (2004b) and depends essentially on the asymptotic geometry of the eigenvalues $\mu_{l,n}$. This distribution may or may not be equal to the $0.5\chi_0^2 : 0.5\chi_1^2$ mixture, depending on the asymptotic behavior of these eigenvalues. A similar result for LRT can be found in Crainiceanu and Ruppert (2004b).

There are several reasons for preferring the distribution in (11) over the $0.5\chi_0^2 : 0.5\chi_1^2$ of Stram and Lee (1994). First, this is the finite sample distribution of the RLRT. Second, the $0.5\chi_0^2 : 0.5\chi_1^2$ asymptotic distribution can be inaccurate when the number of independent sub-vectors of $\boldsymbol{Y}$ is small to moderate or when designs are unbalanced. Typically, the $0.5\chi_0^2 : 0.5\chi_1^2$ provides a conservative approximation of the finite sample distribution with considerable associated losses in power. Third, calculating the distribution in (11) is very fast. Indeed, the distribution in (11) depends only on the eigenvalues $\mu_{l,n}$ of a $K_1 \times K_1$ matrix, which need to be computed only once. Simulation effectively reduces to simulation of $(K_1 + 1)$ χ^2 variables and a grid search over λ. This simulation does not depend on the sample size, n, and is fast (5,000 simulations per second with a 2.66 GHz CPU and 1 Mbyte random access memory). Fourth, when assumptions in Stram and Lee (1994) hold the distribution in (11) converges weakly to the asymptotic $0.5\chi_0^2 : 0.5\chi_1^2$.

5 Linear Mixed Models with Multiple Variance Components

The results in Crainiceanu and Ruppert (2004b) have solved the problem for mixed models with Gaussian outcomes and one variance component. However, in many practical applications there are multiple variance components controlling shrinkage. Two such examples are the onion density and the coronary sinus potassium models in Sects. 1.2 and 1.3, respectively.

The methodology developed by Crainiceanu and Ruppert (2004b) could be used to derive the null distribution for the more general case discussed in this paper. While the result is theoretically interesting, this distribution is obtained by maximizing a stochastic process over the variance components of model (8), which makes the implementation computationally equivalent to the parametric bootstrap. For this reason, Crainiceanu (2003) and Crainiceanu and Ruppert (2004a) suggest using the parametric bootstrap in this context. One could debate the elegance of this approach, but the parametric bootstrap is a practical and robust alternative to the $0.5\chi_0^2 : 0.5\chi_1^2$ approximation.

One problem with the parametric bootstrap is that, in many applications, evaluating the likelihood is computationally expensive and it may not be reasonable to perform thousands of simulations. To illustrate this problem, consider the following simple longitudinal model:

$$Y_{ij} = u_i + f(x_{ij}) + \epsilon_{ij}, \tag{12}$$

where $u_i \sim N(0, \sigma_u^2)$ are random independent subject specific intercepts, $\epsilon_{ij} \sim N(0, \sigma_\epsilon^2)$ are independent errors, $i = 1, \ldots, I$, $j = 1, \ldots, J$, I is the number of subjects and J is the number of observations per subject. Here $f(.)$ is an unspecified population mean function. If the function $f(.)$ is modeled as a linear penalized spline, then testing for linearity of $f(.)$ against a nonparametric alternative is equivalent to testing

$$H_0 : \sigma_b^2 = 0 \quad \text{vs.} \quad H_A : \sigma_b^2 > 0, \tag{13}$$

where σ_b^2 is a variance component controlling the degree of smoothness of $f(.)$.

Computation times both for LRT and RLRT were very long even for small sample sizes. For example, for six subjects and 50 observations per subject, computation time for 10,000 simulations was 4.5 h for R and 1 h for SAS on a server (Intel Xeon 3 GHz CPU). Additionally, run time increased steeply with both I and J for R. For R significant reduction of computation times could be achieved by interfacing it with C or FORTRAN. SAS is faster with its default convergence criterion, but we found numerical imprecisions, especially when estimating the probability mass at zero. These problems were mitigated when the convergence criterion was more stringent, but was accompanied by an increasing proportion of unsuccessful model fits. For more details see the extensive simulation study in Greven et al. (2008). Needless to say that in more complex models with larger sample sizes the computational burden is even more serious, especially when running several tests or performing simulation studies.

Therefore, for many applications there is a need for fast and accurate approximations of the null finite sample distribution of the RLRT for testing $H_{0,s}$. We describe two such approximations. The first approximation was introduced by Greven et al. (2008), is practically instantaneous, and avoids bootstrap. The second approximation was introduced by Crainiceanu (2003) and Crainiceanu and Ruppert (2004a) and uses a simple parametric approximation that reduces the necessary number of bootstrap samples. In extensive simulation studies, Greven et al. (2008) show that both methods outperform the $0.5\chi_0^2 : 0.5\chi_1^2$ approximation and the parametric bootstrap. The approximation used by standard software is the $0.5\chi_0^2 : 0.5\chi_1^2$ approximation. The necessary regularity conditions for this approximation to be asymptotically valid are independence under null and alternative hypothesis, large number of subvectors, and balanced designs. When these conditions are met both approximated distributions discussed in the following converge weakly to $0.5\chi_0^2 : 0.5\chi_1^2$ distribution. However, when conditions are not met, both approximate distributions agree with each other, are different from the $0.5\chi_0^2 : 0.5\chi_1^2$ distribution, and better fit the finite sample distribution of the RLRT.

5.1 Fast Finite Sample Approximation

The approximation proposed by Greven et al. (2008) is a combination of results in Crainiceanu and Ruppert (2004b) and the pseudo-likelihood estimation idea in Gong and Samaniego (1981). Recall that the pseudo-likelihood function is obtained by plugging in a consistent estimator of the nuisance parameters instead of the nuisance parameters. More precisely, let $L(\boldsymbol{\theta}, \boldsymbol{\phi})$ be the likelihood for independent and identically distributed (i.i.d.) random variables $X_1, \ldots, X_n$, where the likelihood depends on the parameters of interest $\boldsymbol{\theta}$ and on nuisance parameters $\boldsymbol{\phi}$. Assume that $L(.,.)$ is a complicated function of $\boldsymbol{\theta}$ and $\boldsymbol{\phi}$, but simple as a function of $\boldsymbol{\theta}$ alone when $\boldsymbol{\phi}$ is fixed. In this case, pseudo-likelihood replaces $\boldsymbol{\phi}$ by a consistent estimator $\hat{\boldsymbol{\phi}}$ and maximizes $L^*(\boldsymbol{\theta}) = L(\boldsymbol{\theta}, \hat{\boldsymbol{\phi}})$ over $\boldsymbol{\theta}$ to obtain the pseudo-maximum likelihood estimator $\hat{\boldsymbol{\theta}}$. The pseudo-LRT for testing $H_0 : \boldsymbol{\theta} = \boldsymbol{\theta}_0$ is then defined as $\text{LRT}^* = 2\log L^*(\hat{\boldsymbol{\theta}}) - 2\log L^*(\boldsymbol{\theta}_0)$.

In our framework, $\boldsymbol{\theta} = (\sigma_s^2, \boldsymbol{\beta}, \boldsymbol{b}_s)$ could be viewed as the parameters of interest, and the $\boldsymbol{b}_i$, $i \neq s$, as nuisance parameters. If the $\boldsymbol{b}_i$'s were known, the outcome vector could be redefined as $\widetilde{\boldsymbol{Y}} = \boldsymbol{Y} - \sum_{i \neq s} \boldsymbol{Z}_i \boldsymbol{b}_i$ and our model could be reduced accordingly. The idea we transfer from pseudo-likelihood estimation is, that under regularity conditions, the prediction of $\sum_{i \neq s} \boldsymbol{Z}_i \boldsymbol{b}_i$ might be good enough to allow the RLRT null distribution for testing $H_0 : \sigma_s^2 = 0$ to be closely approximated by the RLRT distribution when $\sum_{i \neq s} \boldsymbol{Z}_i \boldsymbol{b}_i$ is known. Thus, Greven et al. (2008) use the following reduced model

$$\begin{cases} \widetilde{\boldsymbol{Y}} = \boldsymbol{X}\boldsymbol{\beta} + \boldsymbol{Z}_s \boldsymbol{b}_s + \boldsymbol{\epsilon}, \\ \boldsymbol{b}_s \sim N(\boldsymbol{0}, \sigma_s^2 \boldsymbol{I}_{K_s}), \\ \boldsymbol{\epsilon} \sim N(\boldsymbol{0}, \sigma_\epsilon^2 \boldsymbol{I}_{K_n}), \end{cases} \tag{14}$$

where notations are similar to notations for model (10). The idea is to calculate the RLRT for testing $H_{0,s} : \sigma_s^2 = 0$ using an LMM of type (8) but use an approximated RLRT null distribution based on testing for zero variance in the LMM (14). The advantage of this approach is that this finite sample distribution can be obtained very easily, as described in Sect. 4.

5.2 Mixture Approximation to the Bootstrap

In some cases, one might still want to use a simple parametric bootstrap to determine the distribution of the (R)LRT. Given the steep computational penalty in many applications, we propose to use a parametric approximation to the (R)LRT distribution. While in the case of i.i.d. data the distribution is asymptotically a $0.5\chi_0^2 : 0.5\chi_1^2$ mixture, Crainiceanu and Ruppert (2004b) showed that for correlated responses and finite sample sizes the distribution can severely deviate from this mixture. We propose to use the following finite sample approximation

$$\text{(R)LRT} \overset{d}{\approx} aUD, \tag{15}$$

where $U \sim \text{Bernoulli}(1-p)$, $D \sim \chi_1^2$, $p = P(U = 0)$, and a are unknown constants, and $\overset{d}{\approx}$ denotes approximate equality in distribution. The parameters of the aUD approximation are estimated using a bootstrap sample that is typically much smaller than the one required to estimate small tail probabilities.

Note that the flexible class of distributions in (15) Contains, as a particular case, the $0.5\chi_0^2 : 0.5\chi_1^2$ distribution with $a = 1$ and $p = 0.5$, and is just as easy to use. As the point mass at zero, p, and the scaling factor, a, are unknown in all other cases, we propose to estimate them from a bootstrap sample. The idea of the parametric approximation is to use the entire bootstrap sample to fit a flexible two parameter family of distributions, thus reducing the necessary number of simulations required for estimating tail quantiles. Greven et al. (2008) show that the approximation (15) generally outperforms the $0.5\chi_0^2 : 0.5\chi_1^2$ approximation. This happens in many applications when the correlation structure imposed by the random effects, $\boldsymbol{b}_i$, cannot be ignored, or when the sample size is small to moderate. Note that both approximations are asymptotically identical to the $0.5\chi_0^2 : 0.5\chi_1^2$ approximation when the assumptions in Self and Liang (1987) and Stram and Lee (1994) hold.

This methodology has been applied by Crainiceanu (2003) and Crainiceanu and Ruppert (2004a). Its behavior has been studied in extensive simulation studies by Greven et al. (2008) in a wide variety of settings indicating excellent agreement with long bootstrap simulations. The main strengths of the method are that it requires few bootstrap samples (100–200), provides a finite sample approximation, and can be applied to data with a moderate number of clusters and unbalanced designs.

6 Revisiting the Applications

In this section, we revisit the applications described in Sect. 1. Table 1 provides the RLRT calculated for each application together with the p-value estimated nonparametrically from 10,000 simulations using the parametric bootstrap. We also report the estimated aUD approximation to the bootstrap.

The first row of Table 1 provides results for testing the null hypothesis of a linear spatial drift against a general alternative in the *Loa loa* application. This is testing

Table 1 RLRT testing for the three examples introduced in Sect. 1

Example	Test	Value	p-Value	aUD
Loa loa	$\sigma_b^2 = 0$	127.47	<0.001	$0.66\chi_0^2 + 0.34(0.91\chi_1^2)$
Onion	$\sigma_b^2 = 0$	1.98	0.048	$0.66\chi_0^2 + 0.34(0.91\chi_1^2)$
Dogs	$\sigma_b^2 = 0$	4.89	0.0057	$0.69\chi_0^2 + 0.31(0.88\chi_1^2)$
Dogs	$\sigma_u^2 = 0$	19.03	<0.001	$0.56\chi_0^2 + 0.44(0.96\chi_1^2)$

whether there remains sizable spatial correlation after controlling for the effects of available covariates. In this case RLRT $= 127.47$, suggesting very strong evidence against the linear spatial trend, irrespective of the particular approximation to the null distribution. In this case, because the alternative model is an LMM with one variance component, we actually have the exact null finite sample distribution. The aUD approximation is still displayed because it provides a compact and accurate summary of the exact distribution.

In many testing examples, decisions are not as easy to make as in the *Loa loa* case. Indeed, the second row presents results for the onion density example. The null hypothesis $\sigma_b^2 = 0$ corresponds to testing for the semiparametric binary offset model (two parallel nonparametric functions) against the discrete by continuous interaction model (two nonparametric functions). The value of RLRT $= 1.96$, is much closer to the decision boundary. In such cases, it is reasonable to invest computational effort to obtain the null finite sample distribution. The aUD approximation to the bootstrap suggests serious differences from the $0.5\chi_0^2 : 0.5\chi_1^2$ distribution. In fact, the p-value based on the aUD approximation was 0.048 compared to 0.081 based on the $0.5\chi_0^2 : 0.5\chi_1^2$ approximation. It could seem strange that the two aUD approximations for the widely different testing problems are identical. This is due to the fact that both distributions depend essentially on a very large leading eigenvalue. The following two examples have different distributions because their eigenvalue structure is different.

The last two rows in Table 1 are dedicated to results for the coronary sinus potassium data. The third row corresponds to testing for a quadratic population curve against a general alternative, while the last row corresponds to testing for homogeneity of dog responses around the nonparametric population curve. Both testing procedures suggest strong evidence against the corresponding null hypotheses.

Results in this section were obtained for a fixed number of knots and choice of knot locations. In simulation studies, Crainiceanu (2003) and Greven et al. (2008) showed that the null distribution and power properties do not change substantially by increasing the number of knots as long as the regression design provides an alternative that is flexible enough to capture the potential complexities of the alternative hypothesis. This results are consistent with the results in Ruppert (2002) who showed that 20 knots are enough to fit most functions that do not exhibit extreme changes in curvature.

7 Discussion

LMMs are used in a wide range of applications such as longitudinal studies, hierarchical models or smoothing. The likelihood ratio testing for zero variance components in mixed models has long been a methodological challenge. Research in the last 20 years combined with recent methodological results and simulation studies have led to a better understanding of the framework. Most importantly, the application of likelihood ratio testing for most LMMs has become possible, if not routine.

While this paper is not aimed at answering all questions, several points should be made clear. First, the χ_1^2 approximation can be applied with the acknowledgement that it may provide an excessively conservative approximation to the null distribution. This is safe when the evidence against the null is overwhelming (see, for example the *Loa loa* and the dog response homogeneity examples). Second, the $0.5\chi_0^2 : 0.5\chi_1^2$ approximation can be applied in many situations, especially when testing for homogeneity of a large number of clusters (in the dogs example there are nine dogs). However, this approximation tends to be conservative and lose power in many applications. Effects are less serious than the ones associated to the use of the χ_1^2 approximation. Thus, a nonsignificant effect using the 0.95 quantile of the $0.5\chi_0^2 : 0.5\chi_1^2$ distribution, could, in fact, be significant using the correct null distribution at the same level (see the onion example). Third, in the case of LMMs with one variance component the finite sample distribution of the RLRT is available and easy to obtain. Fourth, in the case of LMMs with more than one variance components the fast finite sample approximation introduced in Sect. 5.1 performs consistently well. Fifth, the aUD approximation introduced in Sect. 5.2 may reduce simulation times while preserving accuracy by using much smaller bootstrap samples.

A natural question to ask is "What should I do if I have a testing problem for a zero variance component in a Linear Mixed Model?" Of course, there are many answers to this particular question, mainly because there are multiple ways of approaching the problem. One such alternative is to use score tests, which are null-based tests, as described in the chapter by Zhang and Lin (2007) in this book.

However, if one decides to use LRTs the following algorithm-like list can provide guidance:

1. Use Restricted Likelihood Ratio Test (RLRT) instead of Likelihood Ratio Test (LRT). This is due to the tendency of ML to strongly underestimate the variance component that ultimately leads to power losses for LRT.
2. If testing for a variance component in an LMM with one variance component use the exact finite sample distribution in Crainiceanu and Ruppert (2004b). If not then continue to Step 3.
3. If possible, obtain 10,000 parametric bootstraps from the null distribution of the RLRT. Compare this distribution with the $0.5\chi_0^2 : 0.5\chi_1^2$ and aUD approximations. Report results based on bootstrap, $0.5\chi_0^2 : 0.5\chi_1^2$ and aUD approximations.
4. If obtaining 10,000 bootstraps is computationally prohibitive, obtain at least 100–200 bootstrap samples. Then continue as in Step 3.
5. Obtain the finite sample approximation described in Sect. 5.1 and compare it with the other approximations.

The author's point of view is that the *null finite sample distribution* is the relevant distribution and not its asymptotic approximation. An asymptotic approximation is relevant when it provides a previously unknown insight, is much easier to use and closely approximates the null finite distribution. Thus, the asymptotic distribution is not the *right distribution* but is one automatic way to approach a testing problem. Fortunately, the variety of applications and problems continues to raise nonstandard problems.

Acknowledgements Ciprian Crainiceanu's work was supported by NIH Grant AG025553-02 on the Effects of Aging on Sleep Architecture.

References

Chernoff, H. (1954). On the distribution of the likelihood ratio. *Annals of Mathematical Statistics 25*, 573–578

Crainiceanu, C. M. (2003). *Ph.D. Thesis: Nonparametric Likelihood Ratio Testing.* Cornell University

Crainiceanu, C. M. and D. Ruppert (2004a). Likelihood ratio tests for goodness-of-fit of a nonlinear regression model. *Journal of Multivariate Analysis 91*, 3552

Crainiceanu, C. M. and D. Ruppert (2004b). Likelihood ratio tests in linear mixed models with one variance component. *Journal of the Royal Statistical Society, Series B 66*(1), 165–185

Crainiceanu, C. M., D. Ruppert, G. Claeskens, and M. P. Wand (2005). Exact likelihood ratio tests for penalised splines. *Biometrika 92*(1), 91–103

Crainiceanu, C. M., P. J. Diggle, and B. Rowlingson (2007). Bivariate binomial spatial modeling of *loa loa* prevalence in tropical africa, with discussions. *Journal of the American Statistical Association 103*, 21–43

Fields Development Team (2006). *Fields: Tools for Spatial Data.* Boulder, CO: National Center for Atmospheric Research

Gong, G. and F. J. Samaniego (1981). Pseudo maximum likelihood estimation: theory and applications. *The Annals of Statistics 9*(4), 861–869

Greven, S., C. M. Crainiceanu, H. Küechenhoff, and A. Peters (2008). Likelihood ratio testing for zero variance components in linear mixed models. *Journal of Computational and Graphical Statistics*

Grizzle, J. E. and D. M. Allan (1969). Analysis of dose and dose response curves. *Biometrics 25*, 357–381

Kammann, E. and M. P. Wand (2003). Geoadditive models. *Applied Statistics 52*, 1–18

Laird, N. and J. H. Ware (1982). Random-effects models for longitudinal data. *Biometrics 38*, 963–974

Moran, P. A. P. (1971). Maximum likelihood estimators in non-standard-conditions. *Proceedings of the Cambridge Philosophical Society 70*, 441–450

Nychka, D. W. and N. Saltzman (1998). Design of air quality monitoring networks. In D. Nychka, L. Cox, and W. Piegorsch (Eds.), *Case Studies in Environmental Statistics*. Berlin Heidelberg New York: Springer

Ratkowsky, D. A. (1983). *Nonlinear Regression Modeling: A Unified Practical Approach.* New York: Marcel Dekker

Ruppert, D. (2002). Selecting the number of knots for penalized splines. *Journal of Computational and Graphical Statistics 11*(4), 735–757

Ruppert, D., M. P. Wand, and R. J. Carroll (2003). *Semiparametric Regression*. Cambridge: Cambridge University Press

Self, S. G. and K.-Y. Liang (1987). Asymptotic properties of maximum likelihood estimators and likelihood ratio tests under nonstandard conditions. *Journal of the American Statistical Association 82*(398), 605–610

Stram, D. O. and J.-W. Lee (1994). Variance components testing in the longitudinal mixed effects model. *Biometrics 50*(3), 1171–1177

Wang, Y. (1998). Mixed effects smoothing spline analysis of variance. *Journal of Royal Statistical Society, Series B 60*, 159–174

Zhang, D. and X. Lin (2007). Variance component testing in generalized linear mixed models for longitudinal/clustered data and other related topics. In D. Dunson (Ed.), *Model Selection in Linear Mixed Models*. Berlin Heidelberg New York: Springer

Variance Component Testing in Generalized Linear Mixed Models for Longitudinal/Clustered Data and other Related Topics

Daowen Zhang and Xihong Lin

1 Introduction

Linear mixed models (Laird and Ware, 1982) and generalized linear mixed models (GLMMs) (Breslow and Clayton, 1993) have been widely used in many research areas, especially in the area of biomedical research, to analyze longitudinal and clustered data and multiple outcome data. In a mixed effects model, subject-specific random effects are used to explicitly model between-subject variation in the data and often assumed to follow a mean zero parametric distribution, e.g., multivariate normal, that depends on some unknown variance components. A large literature was developed in the last two decades for the estimation of regression coefficients and variance components in mixed effects models. See Diggle et al. (2002) and Verbeke and Molenberghs (2000, 2005) for an overview.

In many situations, however, we are interested in testing whether some of the between-subject variations are absent in a mixed effects model. This is equivalent to testing some variance components equal to zero. However, such a null hypothesis places some variance components on the boundary of the parameter space. Hence the commonly used tests, such as the likelihood ratio, Wald and score tests, do not have the traditional chi-squared distribution. In this chapter, we will review the likelihood ratio test and the score test for testing variance components in GLMMs.

A closely related topic is testing whether a covariate effect in a GLMM can be adequately represented by a polynomial of a certain degree. Using a smoothing spline or penalized spline approach, testing for a polynomial covariate effect is equivalent to testing a zero variance component in an induced GLMM. We will review the likelihood ratio test and the score test for testing a parametric polynomial model versus a smoothing spline model for longitudinal data within the generalized additive mixed models framework (Lin and Zhang, 1999).

D. Zhang
Department of Statistics, North Carolina State University, Raleigh, NC 27695, USA
zhang@stat.ncsu.edu

X. Lin
Department of Biostatistics, Harvard School of Public Health, Boston, MA 02115, USA
xlin@hsph.harvard.edu

D. B. Dunson (ed.) *Random Effect and Latent Variable Model Selection*,
DOI: 10.1007/978-0-387-76721-5,

This chapter is organized as follows. In Sect. 2, we present the model specification of a GLMM and briefly review model estimation and inference procedures. In Sect. 3, we review the likelihood ratio test for variance components in GLMMs and illustrate such tests in several common cases of interest. In Sect. 4, we review the score test for variance components in GLMMs, and compare the performance of the likelihood ratio test with the score test in a simple GLMM. In Sect. 6, we review the likelihood ratio test and the score test for testing a polynomial covariate effect versus a nonparametric smoothing spline model for longitudinal data. We illustrate these tests in Sect. 7 through the application of data from a study of infectious disease in Indonesian children. The chapter ends with a discussion in Sect. 8.

2 Generalized Linear Mixed Models for Longitudinal/Clustered Data

Suppose there are m subjects in the sample. For the ith subject, denote by y_{ij} the response measured for the jth observation, e.g., the jth time point for longitudinal data or the jth outcome for multiple outcome data. Similarly, denote by x_{ij} a $p \times 1$ vector of covariates associated with fixed effects and by z_{ij} a $q \times 1$ vector of covariate values associated with random effects. Given subject-specific random effects b_i, the responses y_{ij} are assumed to be conditionally independent and belong to an exponential family with the conditional mean $\mathrm{E}(y_{ij}|b_i) = \mu_{ij}$ and conditional variance $\mathrm{var}(y_{ij}|b_i) = V(\mu_{ij}) = \phi\omega_{ij}^{-1}v(\mu_{ij})$, where ϕ is a positive dispersion parameter, ω_{ij} is a pre-specified weight such as the binomial denominator when y_{ij} is the proportion of events in binomial sampling, and $v(\cdot)$ is the variance function. A generalized linear mixed model (GLMM) relates the conditional mean μ_{ij} to the covariates x_{ij} and z_{ij} as follows:

$$g(\mu_{ij}) = x_{ij}^T\beta + z_{ij}^T b_i, \tag{1}$$

where $g(\cdot)$ is a strictly increasing link function, β is a $p \times 1$ vector of fixed effects (regression coefficients) of x, and b_i is a $q \times 1$ vector of subject-specific random effects of z. The model specification is completed by the usual assumption that $b_i \sim \mathrm{N}\{0, D(\psi)\}$, where ψ is a $c \times 1$ vector of variance components.

Model (1) includes many popular models for continuous and discrete data as special cases. For example, if the y_{ij} are continuous outcome measurements assumed to have a normal distribution given random effects b_i and the link function is the identity link $g(\mu) = \mu$, then model (1) reduces to the following linear mixed model (Laird and Ware, 1982)

$$y_{ij} = x_{ij}^T\beta + z_{ij}^T b_i + \epsilon_{ij}, \tag{2}$$

where $\epsilon_{ij} \overset{iid}{\sim} \mathrm{N}(0, \phi)$ are residual errors. When the y_{ij} are binary responses, a common choice of the link function is the logit link $g(\mu) = \log\{\mu/(1-\mu)\}$. In this case, model (1) reduces to the following logistic-normal model

$$\text{logit}\{\text{P}(y_{ij} = 1|b_i)\} = x_{ij}^T\beta + z_{ij}^T b_i. \tag{3}$$

The log-likelihood function $\ell(\beta, \psi; y)$ given outcome y under model (1) is

$$\exp\{\ell(\beta, \psi; y)\} \propto |D(\psi)|^{-m/2} \prod_{i=1}^{m} \int \exp\left\{\sum_{j=1}^{n_i} \ell_{ij}(\beta, \psi; y_{ij}|b_i) - \frac{1}{2} b_i^T D^{-1}(\psi) b_i\right\} db_i, \tag{4}$$

where

$$\ell_{ij}(\beta, \psi; y_{ij}|b_i) = \int_{y_{ij}}^{\mu_{ij}} \frac{\omega_{ij}(y_{ij} - u)}{\phi v(u)} du$$

is the conditional log-likelihood of y_{ij} given random effects b_i.

Estimation and inference in model (1) are often hampered by the intractable integrations involved in evaluation of likelihood (4) and have been well developed in the past two decades. Our main focus in this paper is on variance component testing in a GLMM. We hence list here some representative work as references. Zeger and Karim (1991) used a Gibbs sampling approach for model estimation and inference. Breslow and Clayton (1993) approximated the likelihood (4) using Laplace approximation and conducted model estimation and inference by maximizing a penalized quasi-likelihood (PQL). Breslow and Lin (1995) and Lin and Breslow (1996) studied the bias in PQL estimators and developed bias-correction methods. Booth and Hobert (1999) proposed an automated Monte Carlo EM algorithm to maximize the integrated likelihood (4).

As usual, throughout this chapter, we will use X for the design matrix of β and Z the design matrix of b. That is, $X = (X_1^T, X_2^T, ..., X_m^T)^T$ where $X_i = (x_{i1}, x_{i2}, \ldots, x_{in_i})^T$, and $Z = \text{diag}\{Z_1, Z_2, \ldots, Z_m\}$ where $Z_i = (z_{i1}, z_{i2}, \ldots, z_{in_i})^T$.

3 The Likelihood Ratio Test for Variance Components in GLMMs

The specification of the subject-specific random effects b_i in model (1) models the source of between-subject variation in the covariate effects of z, which also determines the within-subject correlation. The magnitude of this between-subject variation/within-subject correlation is captured by the magnitude of the elements of $D(\psi)$. In practice, investigators may be interested to see if there is no between-subject variation in some covariate effects of z. Statistically, it is equivalent to testing some or all of the elements of $D(\psi)$ to be zero.

In a regular hypothesis testing setting, a likelihood ratio test (LRT) is the most commonly used test due to its desirable theoretical properties and the fact that it is easy to construct. Under very general regularity conditions, the LRT statistic asymptotically has a χ^2 null distribution with the degrees of freedom equal to the number of independent parameters being tested under the null hypothesis. However, when the elements of $D(\psi)$ are tested, the null hypothesis usually places some or all of the components of ψ on the boundary of the model parameter space, in which case the LRT statistic does not have the usual χ^2 null distribution.

Denote by $\theta = (\beta^T, \psi^T)^T$, a combined vector of regression and variance–covariance parameters in the model. Self and Liang (1987) formulated the asymptotic null distribution of the LRT statistic $-2\ln\lambda_m$ for testing

$$H_0 : \theta_0 \in \Omega_0 \text{ vs. } H_A : \theta_0 \in \Omega_1 = \Omega \backslash \Omega_0,$$

when the true value θ_0 of θ is possibly on the boundary of the model parameter space Ω. Assume that the parameter spaces Ω_1 under H_A and Ω_0 under H_0 can be approximated at θ_0 by cones C_{Ω_1} and C_{Ω_0}, respectively, with vertex θ_0. Self and Liang (1987) showed that under some regularity conditions the LRT statistic $-2\ln\lambda_m$ asymptotically has the same distribution as

$$\inf_{\theta \in C_{\Omega_0} - \theta_0} \{(U - \theta)^T I(\theta_0)(U - \theta)\} - \inf_{\theta \in C_{\Omega} - \theta_0} \{(U - \theta)^T I(\theta_0)(U - \theta)\}, \quad (5)$$

where C_Ω is the cone approximating Ω with vertex at θ_0, $C_\Omega - \theta_0$ and $C_{\Omega_0} - \theta_0$ are translated cones of C_Ω and C_{Ω_0} such that their vertices are the origin, $I(\theta_0)$ is the (Fisher) information matrix at θ_0, and U is a random vector distributed as $\mathrm{N}\{0, I^{-1}(\theta_0)\}$. Alternatively, Self and Liang (1987) expressed (5) as

$$\inf_{\theta \in \tilde{C}_0} \|\tilde{U} - \theta\|^2 - \inf_{\theta \in \tilde{C}} \|\tilde{U} - \theta\|^2, \quad (6)$$

where $\tilde{C} = \{\tilde{\theta} : \tilde{\theta} = \Lambda^{1/2} Q^T \theta$ for all $\theta \in C_\Omega - \theta_0\}$, $\tilde{C}_0 = \{\tilde{\theta} : \tilde{\theta} = \Lambda^{1/2} Q^T \theta$ for all $\theta \in C_{\Omega_0} - \theta_0\}$, $\tilde{U}$ is a random vector from $\mathrm{N}(0, I)$ and $Q \Lambda Q^T$ is the spectral decomposition of $I(\theta_0)$; that is, $I(\theta_0) = Q \Lambda Q^T$, $QQ^T = I$ and $\Lambda = \mathrm{diag}\{\lambda_i\}$. We can use either (5) or (6) to derive the asymptotic null distribution for the LRT statistic depending on the structure of $I(\theta_0)$.

Stram and Lee (1994) applied the above general results of Self and Liang (1987) to investigate the asymptotic null distribution of LRT statistic $-2\ln\lambda_m$ for testing components of $D(\psi)$ for linear mixed model (2). Since the results of Self and Liang (1987) are for a general parametric model, they are also applicable to GLMM (1) as long as one can maximize the likelihood (4) under the null and alternative hypotheses of interest. Here, we list some cases one commonly encounters in practice. For reviews on LRT for variance components in linear mixed models, see the chapter "Likelihood Ratio Testing for Zero Variance Components in Linear Mixed Models" by Crainiceanu.

Case 1. Assume the dimension q of the random effects is equal to one, that is, $D = d_{11}$, and we are testing $H_0 : d_{11} = 0$ vs. $H_A : d_{11} > 0$. For example, consider the random intercept model $Z_{ij} b_i = b_i$ and $b_i \sim N(0, d_{11})$ in model (1).

In this case, $\theta = (\beta^T, d_{11})^T$ and $C_{\Omega_0} = R^p \times \{0\}$ and $C_{\Omega_1} = R^P \times (0, \infty)$. Decompose U and $I(\theta_0)$ in (5) as $U = (U_1^T, U_2)^T$ and $I(\theta_0) = \{I_{jk}\}$ corresponding to β and d_{11}. Some algebra then shows that

$$\inf_{\theta \in C_{\Omega_0} - \theta_0} \{(U - \theta)^T I(\theta_0)(U - \theta)\} = \tilde{U}_2^2,$$

where $\tilde{U}_2 = (I_{22} - I_{21} I_{11}^{-1} I_{12})^{1/2} U_2$, and

$$\inf_{\theta \in C_{\Omega} - \theta_0} \{(U - \theta)^T I(\theta_0)(U - \theta)\} = \tilde{U}_2^2 I(\tilde{U}_2 \leq 0).$$

Therefore, (5) reduces to $\tilde{U}_2^2 I(\tilde{U}_2 > 0)$. It is easy to see that $\tilde{U}_2 \sim \text{N}(0, 1)$. The asymptotic null distribution of $-2\ln\lambda_m$ (as $m \to \infty$) is then a 50:50 mixture of χ_0^2 and χ_1^2.

Denote the observed LRT statistic by T_{obs}. Then, the level α likelihood ratio test will reject $H_0 : d_{11} = 0$ if $T_{\text{obs}} \geq \chi^2_{2\alpha,1}$, where $\chi^2_{2\alpha,1}$ is the $(1 - 2\alpha)$th quantile of the χ^2 distribution with one degree of freedom. The corresponding p-value is $P[\chi_1^2 \geq T_{\text{obs}}]/2$, half of the p-value if the regular but incorrect χ_1^2 distribution were used.

Case 2. Assume $q = 2$ so that $D = \{d_{ij}\}_{2\times 2}$, and we test $H_0 : d_{11} > 0, d_{12} = d_{22} = 0$ vs. $H_A : D$ is positive definite. As an example, consider the random intercept and slope model $z_{ij}^T b_i = b_{0i} + b_{1i} t_{ij}$, where t_{ij} is the time and b_{0i} and b_{1i} are the subject-specific random intercept and slope in longitudinal data assumed to follow $(b_{0i}, b_{1i}) \sim N\{0, D(\psi)\}$. The foregoing hypothesis tests the random intercept model (H_0) versus the random intercept and slope model (H_1).

In this case, $\theta = (\theta_1^T, \theta_2, \theta_3)^T$ where $\theta_1 = (\beta^T, d_{11})^T$, $\theta_2 = d_{12}$ and $\theta_3 = d_{22}$. Under $H_0 : d_{11} > 0$, the translated approximating cone at θ_0 is $C_{\Omega_0} - \theta_0 = R^{p+1} \times \{0\} \times \{0\}$. Under $H_0 \cup H_A$, $d_{11} > 0$ and D is positive semidefinite. This is equivalent to $d_{11} > 0$ and $d_{22} - d_{11}^{-1} d_{12}^2 \geq 0$. Since the boundary defined by $d_{22} - d_{11}^{-1} d_{12}^2 = 0$ for any given $d_{11} > 0$ is a smooth surface, the translated approximating cone at θ_0 under $H_0 \cup H_A$ is $C_\Omega - \theta_0 = R^{p+1} \times R^1 \times [0, \infty)$. Similar to Case 1, decompose U and $I^{-1}(\theta_0)$ in (5) as $U = (U_1^T, U_2, U_3)^T$ and $I^{-1}(\theta_0) = \{I^{jk}\}$ corresponding to θ_1, θ_2 and θ_3. We can then show that

$$\inf_{\theta \in C_{\Omega_0} - \theta_0} \{(U - \theta)^T I(\theta_0)(U - \theta)\} = [U_2, U_3] \begin{bmatrix} I^{22} & I^{23} \\ I^{32} & I^{33} \end{bmatrix}^{-1} \begin{bmatrix} U_2 \\ U_3 \end{bmatrix}, \tag{7}$$

$$\inf_{\theta \in C_{\Omega} - \theta_0} \{(U - \theta)^T I(\theta_0)(U - \theta)\} = (I^{33})^{-1} U_3^2 I(U_3 \leq 0). \tag{8}$$

Since $(U_1^T, U_2, U_3)^T \sim \text{N}\{0, I^{-1}(\theta_0)\}$, the distribution of the difference between (7) and (8) is a 50:50 mixture of χ_1^2 and χ_2^2.

For a given significance level α, the critical value c_α for the LRT can be solved by the following equation using some statistical software:

$$0.5P[\chi_1^2 \geq c] + 0.5P[\chi_2^2 \geq c] = \alpha.$$

Alternatively, the significance level α can also be compared to the LRT p-value

$$p\text{-value} = 0.5P[\chi_1^2 \geq T_{\text{obs}}] + 0.5P[\chi_2^2 \geq T_{\text{obs}}],$$

where T_{obs} is the observed LRT statistic. This p-value is always smaller than the usual but incorrect p-value $P[\chi_2^2 \geq T_{\text{obs}}]$ in this setting. The decision based on this classical p-value is hence conservative.

Case 3. Assume $q > 2$ and we test the presence of the qth element of the random effects b_i in model (1). Denote $D = \begin{pmatrix} D_{11} & D_{12} \\ D_{21} & D_{22} \end{pmatrix}$, where the dimensions of D_{11}, D_{12}, and D_{21} are $s \times s$, $s \times 1$, and $1 \times s$, respectively ($s = q - 1$), and D_{22} is a scalar. Then statistically, we test H_0 : D_{11} is positive definite, $D_{12} = 0$, $D_{22} = 0$ vs. H_A : D is positive definite.

Denote by θ_1 the combined vector of β and the unique elements of D_{11}, $\theta_2 = D_{12}$, and $\theta_3 = D_{22}$. Under H_0, the translated approximating cone at θ_0 is $C_{\Omega_0} - \theta_0 = R^{p+s(s+1)/2} \times \{0\}^s \times \{0\}$. Under $H_0 \cup H_A$, D_{11} is positive definite and D is positive semidefinite. This is equivalent to D_{11} being positive definite and $D_{22} - D_{12}^T D_{11}^{-1} D_{12} \geq 0$ (Stram and Lee (1994), mistakenly used q constraints). Again, since the boundary defined by $D_{22} - D_{12}^T D_{11}^{-1} D_{12} = 0$ for any given positive definite matrix D_{11} is a smooth surface, the translated approximating cone at θ_0 under $H_0 \cup H_A$ is $C_\Omega - \theta_0 = R^{p+s(s+1)/2} \times R^s \times [0, \infty)$. This case is similar to Case 2 except that U_2 is an $s \times 1$ random vector. Therefore, the asymptotic null distribution of LRT statistic is a 50:50 mixture of χ_s^2 and χ_{s+1}^2. The p-value of the LRT test for given observed LRT statistic T_{obs} is equal to $0.5P[\chi_s^2 \geq T_{\text{obs}}] + 0.5P[\chi_{s+1}^2 \geq T_{\text{obs}}]$, which will be closer to the usual but incorrect p-value $P[\chi_{s+1}^2 \geq T_{\text{obs}}]$ as s becomes larger.

Case 4. Suppose the random effects part $z_{ij}^T b_i$ in model (1) can be decomposed as $z_{ij}^T b_i = z_{1ij}^T b_{1i} + z_{2ij}^T b_{2i}$, where $b_{1i} \sim \text{N}\{0, D_1(\psi_1)\}$, $b_{2i} \sim \text{N}(0, \psi_2 I)$ and we test H_0 : $\psi_2 = 0$, and D_1 is positive definite, versus H_A : $\psi_2 > 0$, and D_1 is positive definite. Denote by θ_1 the combined vector of β and the unique elements of D_1, and $\theta_2 = \psi_2$. Since the true values of the nuisance parameters θ_1 are interior points of the corresponding parameter space, we can apply the result of Case 1 to this case. This implies that the asymptotic null distribution of the LRT statistic is a 50:50 mixture of χ_0^2 and χ_1^2.

Case 5. Suppose $D_1(\psi_1)$ in Case 4 takes the form $\psi_1 I$, and we test H_0 : $\psi_1 = 0$, $\psi_2 = 0$ versus H_A : either $\psi_1 > 0$ or $\psi_2 > 0$. Denote $\theta = (\beta^T, \psi_1, \psi_2)$ with $\theta_1 = \beta$, $\theta_2 = \psi_1$ and $\theta_3 = \psi_2$. Under H_0, the translated approximating cone at θ_0 is $C_{\Omega_0} - \theta_0 = R^p \times \{0\} \times \{0\}$. Under $H_0 \cup H_A$, the translated approximating cone at θ_0 is $C_\Omega - \theta_0 = R^p \times [0, \infty) \times [0, \infty)$.

Decompose U and $I(\theta_0)$ in (5) as $(U_1^T, U_2, U_3)^T$ and $I(\theta_0) = \{I_{ij}\}$ corresponding to θ_1, θ_2 and θ_3, and define matrix $\tilde{I}$ as follows:

$$\tilde{I} = \begin{bmatrix} \tilde{I}_{22} & \tilde{I}_{23} \\ \tilde{I}_{32} & \tilde{I}_{33} \end{bmatrix} = \begin{bmatrix} I_{22} & I_{23} \\ I_{32} & I_{33} \end{bmatrix} - \begin{bmatrix} I_{21} \\ I_{31} \end{bmatrix} I_{11}^{-1} [I_{12}, I_{13}].$$

Then $(U_2, U_3)^T \sim N(0, \tilde{I}^{-1})$. Given θ_2 and θ_3, it can be easily shown that

$$\inf_{\theta_1 \in R^p} (U - \theta)^T I(\theta_0)(U - \theta) = [U_2 - \theta_2, U_3 - \theta_3]\tilde{I} \begin{bmatrix} U_2 - \theta_2 \\ U_3 - \theta_3 \end{bmatrix}$$
$$= (\tilde{U}_2 - \tilde{\theta}_2)^2 + (\tilde{U}_3 - \tilde{\theta}_3)^2,$$

where $(\tilde{U}_2, \tilde{U}_3)^T = \tilde{\Lambda}^{1/2}\tilde{Q}^T(U_2, U_3)^T$, $(\tilde{\theta}_2, \tilde{\theta}_3)^T = \tilde{\Lambda}^{1/2}\tilde{Q}^T(\theta_2, \theta_3)^T$, $\tilde{Q}\tilde{\Lambda}\tilde{Q}^T$ is the spectral decomposition of $\tilde{I}$. Therefore, under H_0, we have

$$\inf_{\theta \in C_{\Omega_0} - \theta_0} (U - \theta)^T I(\theta_0)(U - \theta) = \tilde{U}_2^2 + \tilde{U}_3^2.$$

Denote by φ the angle in the radiant formed by the vectors $\tilde{\Lambda}^{1/2}\tilde{Q}^T(1,0)^T$ and $\tilde{\Lambda}^{1/2}\tilde{Q}^T(0,1)^T$, that is, $\varphi = \cos^{-1}\left(\tilde{I}_{23}/\sqrt{\tilde{I}_{22}\tilde{I}_{33}}\right)$ (Self and Liang (1987), who accidentally used I_{jk}), and set $\xi = \varphi/2\pi$, then

$$\inf_{\theta \in C_{\Omega} - \theta_0} (U - \theta)^T I(\theta_0)(U - \theta) = \begin{cases} \tilde{U}_2^2 + \tilde{U}_3^2 & \text{with probability } \xi \\ \tilde{U}_2^2 & \text{with probability } 0.25 \\ \tilde{U}_3^2 & \text{with probability } 0.25 \\ 0 & \text{with probability } 0.5 - \xi. \end{cases}$$

Therefore, the asymptotic null distribution of the LRT statistic is a mixture of χ_0^2, χ_1^2, and χ_2^2 with mixing probabilities ξ, 0.5, and $0.5 - \xi$. Note that since $\tilde{I}$ is a positive definite matrix, the probability ξ satisfies $0 < \xi < 0.5$. In particular, if $\tilde{I}$ is diagonal, the mixing probabilities are 0.25, 0.5, and 0.25.

The asymptotic null distribution of the LRT statistic is relatively easier to study for the above cases. The structure of the information matrix $I(\theta_0)$ and the approximating cones $C_\Omega - \theta_0$ and $C_{\Omega_0} - \theta_0$ play key roles in deriving the asymptotic null distribution. For more complicated cases of testing variance components, although the asymptotic null distribution of the LRT is generally still a mixture of some chi-squared distributions, it may be too difficult to derive the mixing probabilities. In this case, one may use simulation to calculate the p-value.

4 The Score Test for Variance Components in GLMMs

Conceptually, the LRT test for variance components in GLMMs discussed in Sect. 3 is easy to apply. However, the LRT involves fitting GLMM (1) under H_0 and $H_0 \cup H_A$. For many situations, it is relatively straightforward to fit model (1) under H_0. However, one could often encounter numerical difficulties in fitting the full model (1) under $H_0 \cup H_A$. First, fitting model (1) under $H_0 \cup H_A$ involves higher dimensional integration, thus increasing computational burden. Second, if H_0 is true or approximately true, it is often unstable to fit a more complicated model under $H_0 \cup H_A$ as the parameters used to specify H_0 are estimated close to the boundary. For example, although the Laplace approximation used by Breslow and Clayton (1993) and others is recommended for a GLMM with complex parameter boundary, such approximation may work poorly in such cases (Hsiao, 1997). In this section, we discuss score tests for variance components in model (1). One advantage of using score tests is that we only need to fit model (1) under H_0, often dramatically reducing computational burden. Another advantage is that unlike likelihood ratio tests, score tests only require the specification of the first two moments of random effects and are hence robust to mis-specification of the distribution of random effects (Lin, 1997).

We first review the score test for Case 1 discussed in Sect. 3, that is, we assume that there is only one variance component in model (1) for which we would like to conduct hypothesis testing. A one-sided score test is desirable in this case and can be found in Lin (1997) and Jacqmin-Gadda and Commenges (1995). Zhang (1997) discussed a one-sided score test for testing $H_0 : \psi_2 = 0$ for Case 5 in Sect. 3 for a generalized additive mixed model, which includes model (1) as a special case. Verbeke and Molenberghs (2003) discussed one-sided score tests for linear mixed model (2). Lin (1997) derived score statistics for testing single or multiple variance components in GLMMs and considered simpler two-sided tests. Parallel to likelihood ratio tests, the one-sided score tests follow a mixture of chi-square distribution whose weights could be difficult to calculate when multiple variance components are set to be zero under H_0 as illustrated in Case 5. The two-sided score tests assume the score statistics follow a regular chi-square distribution and hence its p-value can be calculated more easily, especially for multiple variance component tests. The two-sided score test has the correct size under H_0, while its power might be lower than the one-sided score and likelihood ratio tests. See the simulation results for more details.

In Case 1, $\psi = d_{11}$. Assume at the moment that β is known. One can show using L'Hôpital's rule or the Taylor expansion (Lin, 1997) that the score for ψ is

$$U_\psi = \left.\frac{\partial \ell(\beta, \psi; y)}{\partial \psi}\right|_{\psi=0} = \frac{1}{2}\sum_{i=1}^{m}\left[\left\{\sum_{j=1}^{n_i} z_{ij} w_{ij} \delta_{ij}\left(y_{ij} - \mu_{ij}^0\right)\right\}^2 - \sum_{j=1}^{n_i} z_{ij}^2 \left\{w_{ij} + e_{ij}\left(y_{ij} - \mu_{ij}^0\right)\right\}\right], \tag{9}$$

where $w_{ij} = [V(\mu_{ij}^0)\{g'(\mu_{ij}^0)\}^2]^{-1}$, $\delta_{ij} = g'(\mu_{ij}^0)$,

$$e_{ij} = \frac{V'\left(\mu_{ij}^0\right) g'\left(\mu_{ij}^0\right) + V\left(\mu_{ij}^0\right) g''\left(\mu_{ij}^0\right)}{V^2\left(\mu_{ij}^0\right)\left\{g'\left(\mu_{ij}^0\right)\right\}^3},$$

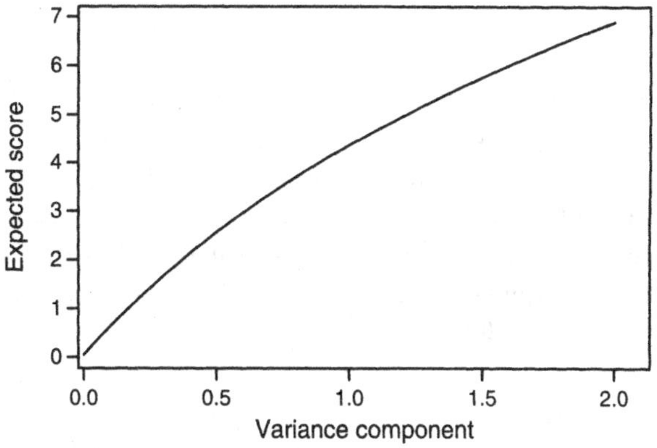

Fig. 1 Expected score as a function of variance component ψ

which is zero for the canonical link function $g(\cdot)$, and μ_{ij}^0 satisfies $g(\mu_{ij}^0) = x_{ij}^T\beta$.

It can be easily shown that the random variable U_ψ defined by (9) has zero mean under $H_0 : \psi = 0$. As argued by Verbeke and Molenberghs (2003), the log-likelihood $\ell(\beta, \psi; y)$ for the linear mixed model (2) on average has a positive slope at $\psi = 0$ when in fact $\psi > 0$. The same argument also applies to GLMM (1). This is because under $H_A : \psi > 0$, the MLE $\widehat{\psi}$ of ψ will be close to ψ so that $\widehat{\psi} > 0$ when the sample size m gets large. If the log-likelihood $\ell(\beta, \psi; y)$ as a function of ψ only is smooth and has a unique MLE $\widehat{\psi}$, which is the case for most GLMMs, the slope U_ψ of $\ell(\beta, \psi; y)$ at $\psi = 0$ will be positive. Indeed, $\mathrm{E}(U_\psi)$ generally is an increasing function of ψ. For example, Fig. 1 plots the expected score $\mathrm{E}(U_\psi)$ vs. ψ for the logistic-normal GLMM (3) where $m = 10, n_i = 5, x_{ij} = 1, \beta = 0.25$, and $z_{ij} = 1$. It is confirmed that $\mathrm{E}(U_\psi)$ increases as ψ increases.

The above argument indicates that a large value of U_ψ provides evidence against $H_0 : \psi = 0$ and we should reject H_0 only if U_ψ is large. Since U_ψ is a sum of independent random variables, classic results show that it will have an asymptotic normal distribution under $H_0 : \psi = 0$ with zero mean and variance equal to $I_{\psi\psi} = \mathrm{E}(U_\psi^2)$, where the expectation is taken at $H_0 : \psi = 0$.

Denote by κ_{rij} the rth cumulant of y_{ij} under H_0. By the properties of the distributions in an exponential family, κ_{3ij} and κ_{4ij} are related to κ_{2ij} via $\kappa_{(r+1)ij} = \kappa_{2ij}\partial\kappa_{rij}/\partial\mu_{ij}$ $(r = 2, 3)$, where $\kappa_{2ij} = \phi\omega_{ij}^{-1}v(\mu_{ij})$ and $\mu_{ij} = \mu_{ij}^0$. Specifically,

$$\kappa_{3ij} = \left(\phi\omega_{ij}^{-1}\right)^2 v'\left(\mu_{ij}\right) v\left(\mu_{ij}\right),$$

$$\kappa_{4ij} = \left(\phi\omega_{ij}^{1}\right)^3 \left[v''\left(\mu_{ij}\right) v(\mu_{ij}) + \{v'(\mu_{ij})\}^2\right] v\left(\mu_{ij}\right).$$

Then $I_{\psi\psi}$ can be shown to be (Lin, 1997)

$$I_{\psi\psi} = \frac{1}{4}\sum_{i=1}^{m}\sum_{j=1}^{n_i} z_{ij}^2 r_{ii},$$

where $r_{ii} = w_{ij}^4 \delta_{ij}^4 \kappa_{4ij} + 2w_{ij}^2 + e_{ij}\kappa_{2ij} - 2w_{ij}^2 \delta_{ij}^2 e_{ij}\kappa_{3ij}$. Therefore, a level α score test for testing $H_0 : \psi = 0$ vs. $H_A : \psi > 0$ will reject $H_0 : \psi = 0$ if $U_\psi \geq z_\alpha I_{\psi\psi}^{1/2}$.

In practice, however, β in U_ψ and $I_{\psi\psi}$ is unknown and has to be estimated under H_0. This is straightforward since under $H_0 : \psi = 0$, GLMM (1) reduces to the standard generalized linear model for independent data $g(\mu_{ij}) = X_{ij}^T \beta$ and existing software can be used to easily calculate the MLE $\widehat{\beta}$ of β under $H_0 : \psi = 0$. In this case, Lin (1997) considered the bias-corrected score statistic to account for the estimation of β under H_0 as

$$U_\psi^c = \left.\frac{\partial \ell(\beta, \psi; y)}{\partial \psi}\right|_{\psi=0,\beta=\widehat{\beta}} = \frac{1}{2}\sum_{i=1}^m \left[\left\{ \sum_{j=1}^{n_i} z_{ij} w_{ij} \delta_{ij} \left(y_{ij} - \widehat{\mu}_{ij}^0 \right) \right\}^2 - \sum_{j=1}^{n_i} z_{ij}^2 w_{0ij} \right], \tag{10}$$

where all quantities are obtained by replacing β by $\widehat{\beta}$, $w_{0ij} = (1 - h_{ij})w_{ij} + e_{ij}(y_{ij} - \widehat{\mu}_{ij}^0)$, and h_{ij} is the corresponding diagonal element of the hat matrix $H = W^{1/2}X(XWX)^{-1}X^T W^{1/2}$, $W = \text{diag}\{w_{ij}\}$, and showed that U_ψ^c has variance

$$\tilde{I}_{\psi\psi} = I_{\psi\psi} - I_{\psi\beta}^T I_{\beta\beta}^{-1} I_{\psi\beta}, \tag{11}$$

where

$$I_{\psi\beta} = \frac{1}{2}\sum_{i=1}^m \sum_{j=1}^{n_i} c_{ij} z_{ij} x_{ij}, \qquad I_{\beta\beta} = X^T W X = \sum_{i=1}^m \sum_{j=1}^{n_i} w_{ij} x_{ij} x_{ij}^T \tag{12}$$

with $c_{ij} = w_{ij}^3 \delta_{ij}^3 \kappa_{3ij} - w_{ij}\delta_{ij} e_{ij} \kappa_{2ij}$. Then the bias-corrected score test at level α would reject H_0 if $T_s = U_\psi^c \geq z_\alpha \tilde{I}_{\psi\psi}^{1/2}$. The one-sided score test presented above is asymptotically equivalent to the likelihood ratio test (Verbeke and Molenberghs, 2003). The two-sided score test assumes the score statistic $T_s = \{U_\psi^c\}^2 / \tilde{I}_{\psi\psi}$ follows a χ^2 distribution. Unlike the regular likelihood ratio test, such a two-sided score test has the correct size under H_0 but is subject to some loss of power. As shown in our simulation studies for a single variance component, the loss of power is minor to moderate for most alternatives. The highest power loss is about 10% when the magnitude of the variance component is moderate.

When the dimension of ψ is greater than 1, suppose we can partition $\psi = (\psi_1, \psi_2)$ where ψ_1 is a $c_1 \times 1$ vector and ψ_2 is a $c_2 \times 1$ vector. We are interested in testing $H_0 : \psi_1 = 0$ vs. $H_A : \psi_1 \geq 0$. Here the inequality is interpreted element-wise. Lin (1997) considered a simple two-sided score test for this multiple variance component test. Specifically, denote by $(\widehat{\beta}, \widehat{\psi}_2)$ the MLE of (β, ψ_2) under $H_0 : \psi_1 = 0$. We can similarly derive the (corrected) score $S_{\psi_1} = m^{-1/2} \partial \ell(\beta, \psi_1; y)/\partial \psi_1|_{\psi_1=0,\beta=\widehat{\beta},\psi_2=\widehat{\psi}_2}$. See Lin (1997) for the special case where each element of ψ represents a variance of a random effect. Asymptotically, S_{ψ_1} has a normal distribution with zero mean and variance equal to the efficient information matrix $H_{\psi_1\psi_1} = m^{-1}\tilde{I}_{\psi_1\psi_1}$ under H_0, where $\tilde{I}_{\psi_1\psi_1}$ is defined similarly to (11) except that $I_{\phi\beta}$ and $I_{\beta\beta}$ are replaced by $I_{\psi_1\gamma}$ and $I_{\gamma\gamma}$ and $\gamma = (\psi_2, \beta)$. The simple two-sided score statistic is defined as

$$T_s = S_{\psi_1}^T H_{\psi_1\psi_1}^{-1} S_{\psi_1} \tag{13}$$

and the p-value is calculated by assuming T_s follows a chi-square distribution with c_1 degrees of freedom.

Silvapulle and Silvapulle (1995) proposed a one-sided score test for a general parametric model and showed that the one-sided score test is asymptotically equivalent to the likelihood ratio test. Verbeke and Molenberghs (2003) extended Silvapulle and Silvapulle (1995) one-sided score test for testing variance components $H_0 : \psi_1 = 0$ vs. $H_A : \psi_1 \in C$ for linear mixed model (2) and showed similar asymptotic equivalence between the one-sided score test and the likelihood ratio test. Hall and Praestgaard (2001) derived a one-sided score test for GLMMs. Then the one-sided score statistic T_s^* is defined as

$$T_s^* = S_{\psi_1}^T H_{\psi_1\psi_1}^{-1} S_{\psi_1} - \inf_{\psi_1 \in C}\{(S_{\psi_1} - \psi_1)^T H_{\psi_1\psi_1}^{-1}(S_{\psi_1} - \psi_1)\}. \tag{14}$$

It is easy to see that T_s^* as defined in (14) has the same asymptotic null distribution of the likelihood ratio test for testing $H_0 : \psi_1 = 0$ vs. $H_A : \psi_1 \in C$. Similarly to the case for the likelihood ratio test, it is critical to determine $H_{\psi\psi}$ and the shape of C, and T_s^* generally follows a mixture of chi-square distributions and we usually have to study the distribution of T_s^* case by case. Both the two-sided test T_s and the one-sided test T_s^* have the correct size under H_0. The two-sided test T_s is much easier to calculate, but is subject to some loss of power. Hall and Praestgaard (2001) conducted extensive simulation studies comparing Lin's (1997) two-sided score test and their one-sided score test for GLMMs with two-dimensional random effects and found similar power loss to the case of a single variance component (Table 4 in Hall and Praestgaard, 2001; the maximum power loss is about 9%).

5 Simulation Study to Compare the Likelihood Ratio Test and the Score Test for Variance Components

We conducted a small simulation study to compare the size and the power of the one-sided and two-sided score tests with the likelihood ratio test. We considered the logistic-normal GLMM (3) by assuming binary responses y_{ij} ($i = 1, 2, \ldots, m = 100$, $j = 1, 2, \ldots, n_i = 5$) were generated from the following logistic-normal GLMM:

$$\text{logitP}(y_{ij} = 1|b_i) = \beta + b_i, \tag{15}$$

where $\beta = 0.25$ and $b_i \sim N(0, \psi)$, with equal spaced ψ in [0,1] by 0.2. For each value of ψ, 500 data sets were generated. The likelihood ratio test described in Sect. 3 and the (corrected) one-sided and two-sided score tests were applied to test

Table 1 Size and power comparisons of the likelihood ratio tests and score tests for a single variance component based on 500 simulations under the logistic model (15)

Method	Size			Power		
	$\psi = 0$	$\psi = 0.2$	$\psi = 0.4$	$\psi = 0.6$	$\psi = 0.8$	$\psi = 1.0$
LRT	0.034	0.370	0.790	0.922	0.990	1.000
Regular LRT	0.020	0.280	0.672	0.882	0.968	0.992
One-sided score test	0.054	0.416	0.834	0.938	0.996	1.000
Two-sided score test	0.050	0.336	0.736	0.910	0.980	0.998

$H_0: \psi = 0$. We compare the performance of the regular but conservative LRT, the appropriate LRT, one-sided and two-sided score test for testing $H_0 : \psi = 0$. The nominal level of all four tests were set at $\alpha = 0.05$.

Table 1 presents the simulation results. The results show that the size of the (correct) likelihood ratio test is little smaller than the nominal level. This is probably due to the numerical instability caused by numerical difficulties in fitting model (15) when in fact there is no random effect in the model, or the fact that the sample size (number of clusters $m = 100$) may not be large enough for the asymptotic theory to take effect. As expected, the regular LRT using χ^2_1 is conservative and the size is too small. On the other hand, both one-sided and two-sided score tests have their sizes very close to the nominal level. The powers of the likelihood ratio test and the one-sided score test are almost the same, although the one-sided score test is slightly more powerful than the LRT, which may be due to the numerical integration required to fit model (15). The two-sided score test has some loss of power compared to the one-sided score test and the correct LRT. However, the p-value of the two-sided score test is much easier to calculate especially for testing for multiple variance components.

6 Polynomial Test in Semiparametric Additive Mixed Models

Lin and Zhang (1999) proposed generalized additive mixed models (GAMMs), an extension of GLMMs where each parametric covariate effect in model (1) is replaced by a smooth but arbitrary nonparametric function, and proposed to estimate each function by a smoothing spline. Using a mixed model representation for a smoothing spline, they cast estimation and inference of GAMMs in a unified framework through a working GLMM, where the inverse of a smoothing parameter is treated as a variance component. A special case of GAMMs is the semiparametric additive mixed models considered by Zhang and Lin (2003)

$$g(\mu_{ij}) = f(t_{ij}) + s_{ij}^T \alpha + z_{ij}^T b_i, \tag{16}$$

where $f(t)$ is an unknown smooth function, i.e., the covariate effect of t is assumed to be nonparametric, s_{ij} some covariate vector, and $b_i \sim N\{0, D(\psi)\}$. For independent (normal) data with the identity link, model (16) reduces to a partially linear model. We are interested in developing a score testing for testing $f(t)$ is a parametric polynomial function versus a smooth nonparametric function. Specifically, we set H_0: $f(t)$ is a polynomial function of degree $K-1$ and H_1: $f(t)$ is a smoothing spline.

Following Zhang and Lin (2003), denote by $t^0 = (t_1^0, t_2^0, \ldots, t_r^0)^T$ a vector of ordered distinct t_{ij}'s and by f a vector of $f(t)$ evaluated at t^0 (without loss of generality, assume $0 < t_1^0 < \cdots < t_r^0 < 1$). The Kth-order ($K \geq 1$) smoothing spline estimator $f(t)$ can be expressed as

$$f(t) = \sum_{k=1}^{K} \delta_k \phi_k(t) + \sum_{l=1}^{r} a_l R(t, t_l^0), \tag{17}$$

where $\{\phi_k(t) = t^{k-1}/(k-1)!\}_{k=1}^K$ is a basis for the space of polynomials of order $K-1$ and $R(t,s)$ is defined as

$$R(t,s) = \frac{1}{\{(K-1)!\}^2} \int_0^1 (s-u)_+^{K-1} (t-u)_+^{K-1} du.$$

Then the smoothing spline estimator of f has the following mixed effect representation:

$$f = T\delta + \Sigma a, \tag{18}$$

where T is an $r \times K$ matrix with the (l,k)th element equal to $\phi_k(t_l^0)$, Σ is a positive matrix with the (l,k)th element equal to $R(t_l^0, t_k^0)$, $\delta = (\delta_1, \ldots, \delta_K)^T$ is a vector of fixed effects and $a = (a_1, a_2, \ldots, a_r)^T \sim \mathrm{N}(0, \tau\Sigma^{-1})$ is a vector of random effects with $\tau \geq 0$ being the inverse of the smoothing parameter for the smoothing spline estimate $f(t)$.

Let $n = \sum_{i=1}^m n_i$ be the total sample size and denote by N the $n \times r$ incidence matrix that maps $\{t_{ij}\}$'s into t^0. Further, denote $S = \{(s_{i1}, \ldots, s_{in_i})^T\}$, $X = (NT, S)$, $B = N\Sigma$, $\mu^b = (\mu_{11}^b, \ldots, \mu_{1n_1}^b, \ldots, \mu_{m1}^b, \ldots, \mu_{mn_m}^b)^T$. Then under the mixed effect representation (18), semiparametric additive mixed model (16) reduces to a GLMM in matrix notation

$$g(\mu) = X\beta + Ba + Zb, \tag{19}$$

where $\beta = (\delta^T, \alpha^T)^T$ are new fixed effects and a and $b = (b_1^T, \ldots, b_m^m)^T$ are independent new random effects. Therefore, the smoothing spline estimator $f(t)$ can be estimated using the estimation procedure for a GLMM, such as maximum penalized quasi-likelihood procedure of Breslow and Clayton (1993).

We are interested in using this spline and mixed model connection to test whether the smoothing spline $f(t)$ in semiparametric additive mixed model (16) can be adequately modeled by a polynomial of order $K-1$, i.e., H_0: $f(t)$ is a polynomial of order $K-1$ and H_A : $f(t)$ is a smoothing spline. From the smoothing spline expression (17), it is clear that $f(t)$ is a polynomial of order $K-1$ if and only if $a_1 = a_2 = \cdots = a_r = 0$. By mixed effect representation (18), this test is equivalent to the variance component test $H_0 : \tau = 0$ vs. $H_A : \tau > 0$. It is hence natural to consider using the variance component likelihood ratio test or score test described in the earlier sections to test $H_0 : \tau = 0$. However, the data do not have independent cluster structure under the alternative $H_A : \tau > 0$. Therefore, the asymptotic null distribution of the likelihood ratio test statistic for testing $H_0 : \tau = 0$ does not follow a 50:50 mixture of χ_0^2 and χ_1^2. In fact, for independent normal data with the identity link, Crainiceanu et al. (2005) showed that, when $f(t)$ is modeled by a penalized spline (similar to a smoothing spline), the LRT statistic asymptotically has approximately 0.95 mass probability at zero. For this special case, Crainiceanu et al. (2005) derived the exact null distribution of the LRT statistic. Their results, however, may not be applicable to testing $H_0 : \tau = 0$ under a more general mixed model representation (19). Furthermore, it could be computationally difficult to calculate this LRT statistic by fitting model (19) under the alternative $H_A : \tau > 0$ as it usually requires high-dimensional numerical integrations.

Due to the special structure of the smoothing matrix Σ, the score statistic of τ evaluated under $H_0 : \tau = 0$ does not have a normal distribution. Zhang and Lin (2003) showed that the score statistic of τ can usually be expressed as a weighted sum of chi-squared random variables with positive but rapidly decaying weights, and its distribution can be adequately approximated by that of a scaled chi-squared random variable.

Under the mixed model representation (19), the marginal likelihood function $L_M(\tau, \psi; y)$ of (τ, ψ) is given by

$$L_M(\tau, \psi; y) \propto |D|^{-m/2} \tau^{-r/2} \int \exp\left\{ \sum_{i=1}^{m} \sum_{j=1}^{n_i} \ell_{ij}(\beta, \psi, b_i; y_{ij}) - \frac{1}{2} \sum_{i=1}^{m} b_i^T D^{-1} b_i - \frac{1}{2\tau} a^T \Sigma a \right\} da\,db\,d\beta. \tag{20}$$

Let $\ell_M(\tau, \psi; y) = \log L_M(\tau, \psi; y)$ be the log-marginal likelihood function of (τ, ψ). Zhang and Lin (2003) showed that the score $U_\tau = \partial \ell_M(\tau, \psi; y)/\partial \tau|_{\tau=0}$ can be approximated by

$$U_\tau \approx \frac{1}{2} \left\{ (Y - X\beta)^T V^{-1} N \Sigma N^T V^{-1} (Y - X\beta) - \mathrm{tr}\left(P N \Sigma N^T \right) \right\} \Big|_{\widehat{\beta}, \widehat{\psi}}, \tag{21}$$

where $\widehat{\beta}$ is the MLE of β and $\widehat{\psi}$ the REML estimate of ψ from the null GLMM (22), and Y is the working vector $Y = X\beta + Zb + \Delta(y - \mu)$ under the null GLMM

$$g(\mu) = X\beta + Zb, \tag{22}$$

where $\Delta = \text{diag}\{g'(\mu_{ij})\}$, $P = V^{-1} - V^{-1}X(X^T V^{-1} X)^{-1} X^T V^{-1}$ and $V = W^{-1} + Z\tilde{D}Z^T$ with $\tilde{D} = \text{diag}\{D, \ldots, D\}$ and W is defined similarly as in Sect. 4 except μ_{ij}^0 is replaced by μ_{ij}. All these matrices are evaluated under the reduced model (22).

Write $U_\tau = \mathcal{U}_\tau - \tilde{e}$, where $\mathcal{U}_\tau$ and $\tilde{e}$ are the first and second terms of U_τ in (21). Zhang and Lin (2003) showed that the mean of $\mathcal{U}_\tau$ is approximately equal to $\tilde{e}$ under $H_0 : \tau = 0$. Similar to the score test derived in Sect. 4, the mean of $\mathcal{U}_\tau$ increases as τ increases. Therefore, we will reject $H_0 : \tau = 0$ when $\mathcal{U}_\tau$ is large, implying a one-sided test. The variance of $\mathcal{U}_\tau$ under H_0 can be approximated by

$$\tilde{I}_{\tau\tau} = I_{\tau\tau} - I_{\tau\psi}^T I_{\psi\psi}^{-1} I_{\tau\psi}, \tag{23}$$

where

$$I_{\tau\tau} = \frac{1}{2}\text{tr}\left\{\left(PN\Sigma N^T\right)^2\right\}, \; I_{\tau\psi} = \frac{1}{2}\text{tr}\left(PN\Sigma N^T P\frac{\partial V}{\partial \psi}\right),$$
$$I_{\psi\psi} = \frac{1}{2}\text{tr}\left(P\frac{\partial V}{\partial \psi}P\frac{\partial V}{\partial \psi}\right). \tag{24}$$

Define $\kappa = \tilde{I}_{\tau\tau}/2\tilde{e}$ and $\nu = 2\tilde{e}^2/\tilde{I}_{\tau\tau}$. Then $S_\tau = \mathcal{U}_\tau/\kappa$ approximately has a χ_ν^2 distribution, and we will reject $H_0 : \tau = 0$ at the significance level α if $S_\tau \geq \chi^2_{\alpha;\nu}$. The simulation conducted by Zhang and Lin (2003) indicates that this modified score test for polynomial covariate effect in the semiparametric additive mixed model (16) has approximately the right size and is powerful to detect alternatives.

7 Application

In this section, we illustrate the likelihood ratio testing and the score testing for variance components in GLMMs discussed in Sects. 3 and 4, as well as the score polynomial covariate effect testing in GAMMs discussed in Sect. 6 through an application to the data from Indonesian children infectious disease study (Zeger and Karim, 1991). Two hundred and seventy-five Indonesian preschool children were examined for up to six quarters for the sign of respiratory infection (0 = no, 1 = yes). Totally there are 2,000 observations in the data set. Available covariates include: age in years, Xerophthelmia status (sign for vitamin A deficiency), gender, height for age, and the presence of stunning and the seasonal sine and cosine. The primary interest of the study is to see if vitamin A deficiency has an effect on the respiratory infection adjusting for other covariates and taking into account the correlation in the data.

Zeger and Karim (1991) used Gibbs' sampling approach to fit the following logistic-normal GLMM

$$\text{logit}(P[y_{ij} = 1|b_i]) = x_{ij}^T\beta + b_i, \tag{25}$$

where y_{ij} is the respiratory infection indicator for the ith child at the jth interview, x_{ij} is the 7×1 vector of the covariates described above with corresponding effects $\beta, b_i \sim \mathrm{N}(0, \theta)$ is the random effect modeling the between-child variation/between-child correlation. No statistically significant effect of vitamin A deficiency on respiratory infection was found.

We can also conduct a likelihood inference for model (25) by evaluating the required integrations using Gaussian quadrature technique. The MLE of θ is $\widehat{\theta} = 0.58$ with $\mathrm{SE}(\widehat{\theta}) = 0.31$, which indicates that there may be between-child variation in the probability of getting respiratory infection. An interesting question is whether we can reject $H_0 : \theta = 0$. The likelihood ratio statistic for this data set is $-2 \ln \lambda_m = 674.872 - 669.670 = 5.2$. The resulting p-value $= 0.5P[\chi_1^2 \geq 5.2] = 0.011$, indicating strong evidence against H_0 using the LRT procedure.

Alternatively, we may apply the score tests to test H_0. The (corrected) score statistic for this data set is 2.678. The p-value from the one-sided score test is 0.0037, and the two-sided score test is 0.0074. Both the tests provide strong evidence against $H_0 : \theta = 0$.

Motivated by their earlier work, Zhang and Lin (2003) considered testing whether $f(\text{age})$ in the following semiparametric additive mixed model can be adequately represented by a quadratic function of age

$$\text{logit}(P[y_{ij} = 1|b_i]) = s_{ij}^T \beta + f(\text{age}_{ij}) + b_i, \tag{26}$$

where s_{ij} are the remaining covariates. The score test statistic described in Sect. 6 for $K = 3$ is $S_\tau = 5.73$ with 1.30 degrees of freedom, indicating a strong evidence against H_0 : $f(\text{age})$ is a quadratic function of age (p-value $= 0.026$). This may imply that nonparametric modeling of $f(\text{age})$ in model (26) is preferred.

8 Discussion

In this chapter, we have reviewed the likelihood ratio test and the score test for testing variance components in GLMMs. The central issue is that the null hypothesis usually places some of the variance components on the boundary of the model parameter space, and therefore the traditional null chi-squared distribution of the LRT statistic no longer applies and the p-value based on traditional LR chi-square distribution is often too conservative. Using the theory developed by Self and Liang (1987), we have reviewed the LRT for some special cases and show the LRT generally follows a mixture of chi-square distribution. To derive the right null distribution of the LRT statistic, one needs to know the (Fisher) information matrix at the true parameter value (under the null hypothesis) and the topological behavior of the neighborhood of the true parameter value. However, as our simulation indicates, the LRT for the variance components in a GLMM may suffer from numerical instability when the variance component is small and numerical integration is high dimensional.

On the other hand, the score statistic only involves parameter estimates under the null hypothesis and hence can be calculated much more straightforward and efficiently. We discussed both the one-sided score test and the much simpler two-sided score test. Both tests have the correct size. The one-sided score test has the same asymptotic distribution as the correct likelihood ratio test. Hence, similar to the LRT, the calculation of the one-sided score test requires the knowledge of the information matrix and the topological behavior of the neighborhood of the true parameter value and also requires computing a mixture of chi-square distributions. The two-sided score test is based on the regular chi-square distribution and has the right size. It is much easier to calculate especially for testing multiple variance components. The simulation studies presented here and in the statistical literature show that the two-sided score test may suffer from some power loss compared to the (correct) likelihood ratio test and the one-sided score test.

We have also reviewed the likelihood ratio test and the score test for testing whether a nonparametric covariate effect in a GAMM can be adequately modeled by a polynomial of certain degree compared to a smoothing spline or a penalized spline function. Although the problem can be reduced to testing a variance component equal to zero using the mixed effects representation of the smoothing (penalized) spline, the GLMM results for likelihood ratio test and the score test for variance components do not apply because the data under mixed effects representation of the spline do not have an independent cluster structure any more. Since the LRT statistic will be prohibitive to calculate for a GLMM with potentially high dimensional random effects, we have particularly reviewed the score test of Zhang and Lin (2003) for testing the parametric covariate model versus the nonparametric covariate model in the presence of a single nonparametric covariate function. Future research is needed to develop simultaneous tests for multiple covariate effects.

References

Booth, J.G. and Hobert, J.P. (1999). Maximizing generalized linear mixed model likelihoods with an automated Monte Carlo EM algorithm. *Journal of the Royal Statistical Society, Series B* **61**, 265–285

Breslow, N.E. and Clayton, D.G. (1993). Approximate inference in generalized linear mixed models. *Journal of the American Statistical Association* **88**, 9–25

Breslow, N.E. and Lin, X. (1995). Bias correction in generalized linear mixed models with a single component of dispersion. *Biometrika* **82**, 81–91

Crainiceanu, C., Ruppert, D., Claeskens, G. and Wand, M.P. (2005). Exact likelihood ratio tests for penalized splines. *Biometrika* **92**, 91–103

Diggle, P.J., Heagerty, P., Liang, K.Y. and Zeger, S.L. (2002). *Analysis of Longitudinal Data*. Oxford University Press, Oxford

Hall, D.B. and Praestgaard, J.T. (2001). Order-restricted score tests for homogeneity in generalized linear and nonlinear mixed models. *Biometrika* **88**, 739–751

Hsiao, C.K. (1997). Approximate Bayes factors when a model occurs on the boundary. *Journal of the American Statistical Association* **92**, 656–663

Jacqmin-Gadda, H. and Commenges, D. (1995). Tests of homogeneity for generalized linear models. *Journal of the American Statistical Association* **90**, 1237–1246

Laird, N.M. and Ware, J.H. (1982). Random effects models for longitudinal data. *Biometrics* **38**, 963–974

Lin, X. (1997). Variance component testing in generalized linear models with random effects. *Biometrika* **84**, 309–326

Lin, X. and Breslow, N.E. (1996). Bias correction in generalized linear mixed models with multiple components of dispersion. *Journal of the American Statistical Association* **91**, 1007–1016

Lin, X. and Zhang, D. (1999). Inference in generalized additive mixed models using smoothing splines. *Journal of the Royal Statistical Society, Series B* **61**, 381–400

Self, S.G. and Liang, K.Y. (1987). Asymptotic properties of maximum likelihood estimators and likelihood ratio tests under nonstandard conditions. *Journal of the American Statistical Association* **82**, 605–610

Silvapulle, M.J. and Silvapulle, P. (1995). A score test against one-sided alternatives. *Journal of the American Statistical Association* **90**, 342–349

Stram, D.O. and Lee, J.W. (1994). Variance components testing in the longitudinal mixed effects model. *Biometrics*, **50**, 1171–1177

Verbeke, G. and Molenberghs, G. (2000). *Linear Mixed Models for Longitudinal Data.* Springer, Berlin Heidelberg New York

Verbeke, G. and Molenberghs, G. (2003). The use of score tests for inference on variance components. *Biometrics* **59**, 254–262

Verbeke, G. and Molenberghs, G. (2005). *Models for Discrete Longitudinal Data.* Springer, Berlin Heidelberg New York

Zeger, S.L. and Karim, M.R. (1991). Generalized linear models with random effects; a Gibbs sampling approach. *Journal of the American Statistical Association* **86**, 79–86

Zhang, D. (1997). *Generalized Additive Mixed Model*, unpublished Ph.D. dissertation, Department of Biostatistics, University of Michigan

Zhang, D. and Lin, X. (2003). Hypothesis testing in semiparametric additive mixed models. *Biostatistics* **4**, 57–74

Bayesian Model Uncertainty in Mixed Effects Models

Satkartar K. Kinney and David B. Dunson

1 Introduction

1.1 Motivation

Random effects models are widely used in analyzing dependent data, which are collected routinely in a broad variety of application areas. For example, longitudinal studies collect repeated observations for each study subject, while multi-center studies collect data for patients nested within study centers. In such settings, it is natural to suppose that dependence arises due to the impact of important unmeasured predictors that may interact with measured predictors. This viewpoint naturally leads to random effects models in which the regression coefficients vary across the different subjects. In this chapter, we use the term "subject" broadly to refer to the independent experimental units. For example, in longitudinal studies, the subjects are the individuals under study, while in multi-center studies the subjects correspond to the study centers.

In applications of random effects models, one is typically faced with uncertainty in the predictors to be included in the fixed and random effects components of the model. The predictors included in the fixed effects component are correlated with the population-averaged response, while the predictors included in the random effects component have varying coefficients for the different subjects. This variability in the coefficients induces a predictor-dependent correlation structure in the repeated observations upon marginalizing out the random effects. For fixed effects models, there is a rich literature on methods for subset selection and inferences from both frequentist and Bayesian perspectives; however, subset selection for the random effects component has received limited attention.

S.K. Kinney
National Institute of Statistical Sciences
saki@niss.org

D.B. Dunson
Department of Statistical Science, Duke University
dunson@stat.duke.edu

D. B. Dunson (ed.) *Random Effect and Latent Variable Model Selection*,
DOI: 10.1007/978-0-387-76721-5,

One reason for the limited attention to this problem is the common perspective that the primary focus of inference is the fixed effects component of the model, while the dependence structure is merely a "nuisance." From this viewpoint, a relatively simple model for the random effects component, such as a random intercept model, is thought to be sufficient to account for within-subject dependence. There are a few problems with this paradigm. First, it is seldom the case that scientific interest lies only in the predictor for a hypothetical "typical" subject having average random effect values. In clinical trials, for example, variability among the subjects is also important. If the impact of a drug therapy varies considerably among different individuals, this suggests that efficacy is higher for certain subgroups, a finding with considerable clinical implications. Second, one may obtain invalid inferences on the fixed effect coefficients if the random effects component is misspecified.

The focus of this chapter is on applying Bayesian methods for model uncertainty to the random effects subset selection problem. Our goal is to provide a practically-motivated background on the relevant literature, with a focus on the methodology proposed by Kinney and Dunson (2007). The hope is that this tutorial will motivate increased use of Bayesian methods in this area, while also stimulating new research. We begin with a brief overview of the literature on model selection and inferences on variance components in random effects models.

1.2 Frequentist Literature

One of the difficulties in generalizing methods for subset selection and inferences in linear regression models (see, for example, Mitchell and Beauchamp, 1988) to the random effects setting is that the likelihood cannot, in general, be obtained analytically. This is because the likelihood is specified as an integral of a conditional likelihood given the random effects over the random effects distribution, with this integral typically not available in closed form. Motivated by this problem, there is a rich literature on approximations to the marginal likelihood obtained by integrating out the random effects. Sinharay and Stern (2001) summarize the major approaches, including marginal maximum likelihood, restricted maximum likelihood, and quasilikelihood. The marginal maximum likelihood approach evaluates the likelihood using quadrature or a Laplace approximation and computes maximum likelihood estimates of model parameters using traditional numeric optimization approaches. This approach tends to underestimate variance parameters; hence, Stiratelli et al. (1984) suggest an approximate E-M algorithm for computing the restricted maximum likelihood estimate of the variance matrix. An alternative from Breslow and Clayton (1993) is the quasilikelihood method.

After obtaining an accurate approximation to the likelihood, likelihood ratio test statistics can be computed to compare nested random effects models; however, when the nested models differ in the random effects that are included, the typical likelihood ratio test asymptotic theory does not apply since the null hypothesis lies at the boundary of the parameter space. Specifically, the null hypothesis corresponds to setting one or more of the random effects variances equal to zero, with these parameters restricted to be positive under the alternative. Potentially, to avoid relying on knowledge of the exact or asymptotic distribution of the likelihood ratio test statistic (discussed in the chapter by Ciprian Crainiceanu), one could apply a parametric bootstrap (Sinharay and Stern, 2001). As an alternative to a likelihood ratio test, one could consider a score test, such as that considered by Lin (1997) for testing whether all the variance components in a GLMM are zero (see also Verbeke and Molenberghs, 2003; Hall and Praestgaard, 2001 and the chapter by Daowen Zhang and Xihong Lin).

Even if one can obtain accurate p-values for likelihood ratio tests or score tests comparing nested random effects models, it is not clear how to use such methods to appropriately account for uncertainty in subset selection. Potentially, one can apply a stepwise procedure, but the model selected may be sensitive to the order in which predictors are added and the level of p-value cutoff for inclusion or exclusion. In addition, unless one accounts for uncertainty in the selection process, it is not appropriate to base inferences on the estimates of the coefficients under the final selected model. One can potentially address this concern by selecting the model on a training subset of the data, though it is not clear how to optimally choose the training and test samples.

As an alternative approach, which addresses some of these concerns, Jiang et al. (2008) proposed an innovative "fence" method. The fence method gives a single subset of predictors to include in the random effects component, but does not provide a measure of uncertainty in the subset selection process or allow inferences on whether a given predictor has a random coefficient. In addition, if predictions are of interest, one can obtain more realistic measures of uncertainty and potentially more accurate predictions by allowing for errors in model selection. This is particularly important when there are many predictors, because in such cases any single selected model may not be markedly better than all of the competing models.

1.3 Bayesian Approach

Given the practical difficulties that arise in implementing a frequentist approach to this problem, we focus on Bayesian methods. Advantages of the Bayesian approach include (1) lack of reliance on asymptotic approximations to the marginal likelihood obtained by integrating out the random effects or to the distribution of the likelihood ratio test statistic; (2) ability to fully account for model uncertainty through a probabilistic framework that assigns each model in the list prior and posterior probabilities; and (3) allowance for the incorporation of prior information. Practical disadvantages include ease of implementation given the lack of procedures in standard software packages, computational intensity, and sensitivity to prior distributions. These concerns are likely to decrease in the coming years, with new procedures for implementing Bayesian analyses in SAS in a computationally efficient manner and with ongoing research in default priors for model selection in hierarchical models.

We review Bayesian model uncertainty in general in Sect. 2 and in the context of mixed models in Sect. 3. Section 4 describes a Bayesian approach for linear mixed models and discusses prior specification. A modification for binary logistic models is outlined in Sect. 5. Section 6 provides a simulation example and Sect. 7 a data example. Additional extensions are discussed in Sect. 8 and concluding remarks are given in Sect. 9.

2 Bayesian Model Uncertainty

2.1 Subset Selection in Linear Regression

Let us first consider a normal linear model $y = X\boldsymbol{\beta} + \epsilon, \epsilon \sim N(0, \sigma^2)$ with no random effects. From the Bayesian perspective, the model parameters are considered random variables with probability distributions. When fitting the model, prior distributions are assigned to each parameter, and posterior distributions are obtained by updating the prior with the information in the likelihood. Unless conjugate priors are used, the posterior distributions are not available in closed analytic form; hence, Markov chain Monte Carlo (MCMC) algorithms are typically used to produce autocorrelated draws from the joint posterior distribution of the parameters. After convergence of the MCMC chain, the draws can be used to estimate posterior summaries. When performing posterior computation for a single model with no model uncertainty, typical posterior summaries include posterior means, standard deviations, and credible intervals.

In the Bayesian paradigm, model uncertainty can be addressed simultaneously with parameter uncertainty by placing priors $p(M_k)$ on each possible model $M_1, \ldots, M_K$ in addition to the model parameters $p(\boldsymbol{\beta}|M_k, \sigma^2)$ and $p(\sigma^2)$. For example, in normal linear regression analyses, it is common to have uncertainty in the subset of predictors to be included in the regression model. If there are p candidate predictors, then there are 2^p possible subsets, with each M_k corresponding to a different subset. In this case, $p(M_k)$ is the prior probability of subset M_k, which is commonly chosen to be uniform or based on the size of the model. If one allows predictors to be included independently with 0.5 prior probability, then $p(M_k) = \binom{p}{\#M_k} 0.5^p$, where $\#M_k$ is the number of predictors included in model M_k. For normal linear regression models, one can choose a conjugate normal inverse-gamma prior. For example, g-priors or mixtures of g-priors (Liang et al., 2005) are commonly-used.

The posterior model probabilities can be calculated using Bayes rule as follows:

$$p(M_k|y) = \frac{p(y|M_k)p(M_k)}{\sum_k p(y|M_k)p(M_k)},$$

where

$$p(y|M_k) = \int p(y|\boldsymbol{\beta}, \sigma^2, M_k)p(\boldsymbol{\beta}|\sigma^2, M_k)p(\sigma^2)d\boldsymbol{\beta}, d\sigma^2$$

is the marginal likelihood of the data under model M_k. This marginal likelihood is available analytically for normal linear regression models when conjugate normal inverse-gamma priors are chosen for $(\boldsymbol{\beta}, \sigma^2)$; however, in generalized linear models and in normal linear models with random effects, the marginal likelihood will not be available analytically. In such cases, it is common to rely on the Laplace approximation, or to use simulation-based approaches to approximate the marginal likelihood and/or posterior model probabilities.

Even in linear regression models in which one can obtain the exact marginal likelihood for any particular model M_k, calculation of the exact posterior model probabilities may not be possible when the number of models is very large. For example, in the subset selection problem, the number of models in the list is 2^p, which grows very rapidly with p so that one cannot calculate the marginal likelihoods for all models in the list even for moderate p. This problem has motivated a literature on stochastic search variable selection (SSVS) algorithms, which use MCMC methods to explore the high-dimensional model space in an attempt to rapidly identify high posterior probability models (George and McCulloch, 1993, 1997).

The SSVS algorithm of George and McCulloch (1997) uses a Gibbs sampler (Gelfand and Smith, 1990) to search for models having high posterior probability by embedding all the models in the full model. This is accomplished by choosing a prior for the regression coefficients that is a mixture of a continuous density and a component concentrated at zero. Because predictors having zero or near-zero coefficients effectively drop out of the model, the component concentrated at zero allocates probability to models having one or more predictors excluded. Such component mixture priors are commonly referred to as variable selection or point mass mixture priors. They are very convenient computationally, as they allow one to run a single Gibbs sampler as if doing computation under the full model. By randomly switching from the component concentrated at zero to the more diffuse component, the chain effectively moves between models corresponding to different subsets of the predictors being selected.

After discarding initial burn-in draws, one can estimate the posterior model probabilities using the proportion of MCMC draws spent in each model. In general, all 2^p models will not be visited; hence, many or most of the candidate models will be estimated to have zero posterior probability. Although there is no guarantee that the model with the highest posterior probability will be visited, when p is large, SSVS tends to quickly locate good models. Model-averaged estimates may also be obtained for model coefficients by averaging the parameter estimates over all MCMC draws. Marginal inclusion probabilities for each predictor are estimated by the proportion of draws spent in models containing that predictor.

Because posterior probabilities for any specific model tend to be very small in large model spaces, it may be unreliable to base inferences on any one selected model. This problem is not unique to Bayesian model selection, as other model selection criteria (e.g., AIC) may have similar values for many models when the number of candidate models is large. An advantage of the Bayesian approach is that it provides a well calibrated and easily interpretable score for each model. Hence, one can consider a list of the top 10 or 100 models, examining the size of the posterior probabilities allocated to each of these models. Such an exercise provides a much more realistic judge of uncertainty than approaches that seek to identify a single best model based on some criteria.

Marginal inclusion probabilities provide a measure of the weight of evidence that a particular predictor should be included in the model, adjusting for uncertainty in the other predictors that are included. For example, suppose the first candidate predictor (age) is included with posterior probability of 0.98. Then one has strong evidence that age is an important predictor. However, if the marginal inclusion probability was instead 0.02, one would have evidence that age does not need to be included. Posterior inclusion probabilities that are not close to 0 or 1 are less conclusive.

When the focus is on prediction instead of inferences, one can use Bayesian model averaging, which is performed by weighting model-specific Bayesian predictions by the posterior model probabilities. This approach has advantages over classical methods, which instead rely on a single selected model, ignoring uncertainty in selecting that model. For a detailed review of Bayesian model averaging and selection, refer to Clyde and George (2004).

2.2 Bayes Factors and Default Priors

Bayes factors provide a standard Bayesian weight of evidence in the data in favor of one model over another. The Bayes factor in favor of model M_1 over model M_0 is defined as the posterior odds of M_1 divided by the prior odds of M_1

$$\mathrm{BF}_{10} = \frac{p(M_1 \mid y)}{p(M_0 \mid y)} \times \frac{p(M_0)}{p(M_1)} = \frac{p(y|M_1)}{p(y|M_0)},$$

which is simply the ratio of marginal likelihoods under the two different models. Unlike frequentist testing based on p-values, Bayes factors have the advantage of treating the two competing models (say, null and alternative) symmetrically, so that one obtains a measure of support in the data for a model, which is appropriate regardless of whether the models are nested. Hence, we do not obtain a test for whether the large model is "significantly" better, but instead rely on the intrinsic Bayes penalty for model complexity to allow coherent comparisons of non-nested models of different sizes.

A potential drawback (or advantage in certain settings) is that the Bayes factor has a well-known sensitivity to the prior, and improper priors cannot be chosen. This restriction does not hold if one wishes to do inferences under a single model, as long as the posterior is proper. However, in conducting model comparisons, the Bayes factor is only defined up to an arbitrary constant that depends on the variance of the prior. As the prior variance increases, there is an increasing tendency to favor smaller models. Hence, it is important to either choose an informative prior based on subject matter knowledge or to choose a proper default prior, chosen to yield good Bayesian and/or frequentist properties. In subset selection for normal linear regression models, the Zellner–Siow prior (Zellner and Siow, 1980) is a commonly-used default, with recent work proposing alternative mixtures of g-priors (Liang et al., 2005).

The popular Bayesian information criterion (BIC) was originally derived by Schwarz (1978) starting with a Laplace approximation to the marginal likelihood, and making some simplifying assumptions, including the use of a unit information prior. For normal linear regression models, the unit information prior corresponds to a special case of the Zellner g-prior in which $g = n$, so that the amount of information in the prior is equivalent to one observation. Model selection via the BIC closely approximates model selection based on Bayes factors under a wide range of problems for a particular type of default prior (Raftery, 1995). However, for hierarchical models and models in which the number of parameters increases with sample size, the BIC is not justified (Pauler et al., 1999, Berger et al., 2003).

3 Bayesian Subset Selection for Mixed Effects Models

In contrast to the rich literature on Bayesian subset selection for fixed effects, there is very little work on selection of random effects. Pauler et al. (1999) compare variance component models using Bayes factors and Sinharay and Stern (2001) consider the problem of comparing two GLMMs using the Bayes factor. Motivated by sensitivity to the choice of prior, Chung and Dey (2002) develop an intrinsic Bayes factor approach for balanced variance component models. Chen and Dunson (2003) developed a stochastic search variable selection (SSVS) (George and McCulloch, 1993; Geweke, 1996) approach for fixed and random effects selection in the linear mixed effects model. Relying on Taylor series approximation to intractable integrals, Cai and Dunson (2006) recently extended this approach to all GLMMs (refer to chapter by Cai and Dunson for further details).

3.1 Bayes Factor Approximations

The BIC is not appropriate for comparing models with differing numbers of random effects, as the required regularity conditions are not met when the null hypothesis corresponds to a parameter falling on the boundary of the parameter space (Pauler et al., 1999). Several Bayes factor approximations for testing variance components are reviewed in Sinharay and Stern (2001). Most of these involve estimation of $p(y|M_1)$ and $p(y|M_0)$ to obtain the Bayes factor. A modification to the Laplace approximation which accommodates the boundary case is applied by Pauler et al. (1999). As calculation of $p(y|M)$ involves solving an integral that is often not available analytically, one can apply standard approximations such as quadrature and importance sampling.

A practical issue with importance sampling is the selection of the target distribution. Meng and Wong (1996) extend the importance of the sampler idea and suggest a bridge sampling approach for approximating $p(\mathbf{y}|M)$. An MCMC algorithm using Gibbs sampling was developed by Chib (1995). A harmonic estimator, consistent for simulations though otherwise unstable, is proposed by Newton and Raftery (1994). Lastly, an approach suggested by Green (1995) is described which computes the Bayes factors directly using a reversible-jump MCMC algorithm which can move between models with parameter spaces of differing dimension. This is likely to be computationally intensive, and in Sinharay and Stern (2001) it was the slowest approach, whereas the Laplace approximation was the fastest.

3.2 Stochastic Search Variable Selection

In extending Bayesian model selection procedures for linear models to linear mixed effects models, the two primary considerations are the prior specification and posterior computation. The structure of the random effects covariance matrix needs to be considered, and the model parameterizations and prior structure carefully chosen so that the MCMC algorithm may move between models with both differing fixed effects and random effects. The efficiency of the posterior computation also needs to be considered; algorithms that explore the model space efficiently and quickly locate areas of high posterior probability are needed.

As described in Sect. 2, stochastic search variable selection (SSVS) is a promising approach for Bayesian model uncertainty using Gibbs sampling. The SSVS approach has been applied successfully in a wide variety of regression applications, including challenging gene selection problems. One challenge in developing SSVS approaches for random effects models is the constraint that the random effects covariance matrix $\mathbf{\Omega}$ be positive semi-definite. Chen and Dunson (2003) addressed this problem by using a modified Cholesky decomposition of $\mathbf{\Omega}$

$$\mathbf{\Omega} = \mathbf{\Lambda}\mathbf{\Gamma}\mathbf{\Gamma}'\mathbf{\Lambda}, \tag{1}$$

where $\mathbf{\Lambda}$ is a positive diagonal matrix with diagonal elements $\boldsymbol{\lambda} = (\lambda_1, \ldots, \lambda_q)'$ proportional to the random effects standard deviations, so that setting $\lambda_l = 0$ is equivalent to dropping the lth random effect from the model, and $\mathbf{\Gamma}$ is a lower triangular matrix with diagonal elements equal to 1 and free elements that describe the random effects correlations. In the case of independent random effects, $\mathbf{\Gamma}$ is simply the identity matrix $\boldsymbol{I}$ and the diagonal elements $\lambda_l, l = 1, \ldots, q$ of $\mathbf{\Lambda}$ equal the random effects standard deviations.

In the next section, we revisit the SSVS approach of Chen and Dunson (2003) for linear mixed models, with additional consideration given to the prior structure and posterior computation. We will then discuss an extension to logistic models.

4 Linear Mixed Models

If we have n subjects under study, each with n_i observations, $i = 1, \ldots, n$, let y_{ij} denote the jth response for subject i, $\boldsymbol{X}_{ij}$ a $p \times 1$ vector of predictors, and $\boldsymbol{Z}_{ij}$ a $q \times 1$ vector of predictors. Then the linear mixed effects (LME) model is denoted as

$$y_{ij} = \boldsymbol{X}'_{ij}\boldsymbol{\beta} + \boldsymbol{Z}'_{ij}\boldsymbol{a}_i + \epsilon_{ij}, \quad \epsilon_{ij} \sim N(0, \sigma^2), \tag{2}$$

where $\boldsymbol{a}_i \sim N(\boldsymbol{0}, \boldsymbol{\Omega})$. Here $\boldsymbol{\beta} = (\beta_1, \ldots, \beta_p)'$ are the fixed effects and $\boldsymbol{a}_i = (a_{i1}, \ldots, a_{iq})'$ are the random effects. In practice $\boldsymbol{Z}_{ij}$ is typically chosen to be a subset of the predictors in $\boldsymbol{X}_{ij}$ believed to have random effects, often only the intercept for simplicity. If we let $\boldsymbol{X}_{ij}$ and $\boldsymbol{Z}_{ij}$ include all candidate predictors, then the problem of interest is to locate a subset of these predictors to be included in the model.

With the help of covariance decomposition in (1) we can use SSVS, and write (2) as

$$y_{ij} = \boldsymbol{X}'_{ij}\boldsymbol{\beta} + \boldsymbol{Z}'_{ij}\boldsymbol{\Lambda}\boldsymbol{\Gamma}\boldsymbol{b}_i + \epsilon_{ij}, \quad \epsilon_{ij} \sim N(0, \sigma^2), \tag{3}$$

where $\boldsymbol{b}_i \sim N(\boldsymbol{0}, \boldsymbol{I})$. Chen and Dunson (2003) show that by rearranging terms, the diagonal elements, $\lambda_l, l = 1, \ldots, q$, of $\boldsymbol{\Lambda}$ can be expressed as linear regression coefficients, conditional on $\boldsymbol{\Gamma}$ and $\boldsymbol{b}_i$. Similarly, the free elements $\gamma_k, k = 1, \ldots, q(q-1)/2$, of $\boldsymbol{\Gamma}$ can be expressed as linear regression coefficients, conditional on $\boldsymbol{\Lambda}$ and $\boldsymbol{b}_i$. Hence the variance parameters $\boldsymbol{\lambda}$ and $\boldsymbol{\gamma}$ have desirable conditional conjugacy properties for constructing a Gibbs sampling algorithm for sampling the posterior distribution and we are able to use the SSVS approach.

4.1 Priors

Prior selection is a key step in any Bayesian analysis; however, in this context it is particularly important as problems can arise when default priors are applied without caution. In particular, flat or excessively diffuse priors are not recommended for hierarchical models given the potential for an improper posterior and the difficulty of verifying propriety due to the intractable nature of the density, even when the output from a Gibbs chain seems reasonable (Hobert and Casella, 1996). Proper distributions are also desired for Bayes factors to be well-defined (Pauler et al., 1999). The arbitrary multiplicative constants from improper priors carry over to the marginal likelihood $p(\boldsymbol{y}|M)$ resulting in indeterminate model probabilities and Bayes factors (Berger and Pericchi, 2001).

A mixture of a point mass at zero and a normal or heavier-tailed distribution is a common choice of prior for fixed effects coefficients, $\beta_l, l = 1, \ldots, p$, in Bayesian model selection problems. Smith and Kohn (1996) introduce a vector $\boldsymbol{J}$ of indicator variables, where $J_l = 1$ indicates that the lth variable is in the model, $l = 1, \ldots, p$, and assign a Zellner g-prior (Zellner and Siow, 1980) to $\boldsymbol{\beta}_{\boldsymbol{J}}$, the vector of coefficients in the current model. As a notational convention, we let $\boldsymbol{\beta}$ denote the $p \times 1$ vector $(\{\beta_l : J_l = 1\} = \boldsymbol{\beta}_{\boldsymbol{J}}, \{\beta_l : J_l = 0\} = \mathbf{0})$. Hence, conditional on the model index $\boldsymbol{J}$, the prior for $\boldsymbol{\beta}$ is induced through the prior for $\boldsymbol{\beta}_{\boldsymbol{J}}$.

Consistency issues can arise when comparing models based on these priors; however, for linear models, placing a conjugate gamma prior on g induces a t prior on the coefficients. In the special case where the t distribution has degrees of freedom equal 1, the Cauchy distribution is induced, which has been recommended for Bayesian robustness (Clyde and George, 2004). This can be considered a special case of mixtures of g-priors, proposed by Liang et al. (2005) as an attractive computational solution to the consistency and robustness issues with g-priors, and an alternative to the Cauchy prior, which does not yield a closed-form expression for the marginal likelihood. As choosing g can affect model selection, with large values concentrating the prior on small models with a few large coefficients and small values of g concentrating the prior on saturated models with small coefficients, several approaches for handling g have been proposed (Liang et al., 2005). Recommendations include the unit information prior (Kass and Wasserman, 1995), which in the normal regression case corresponds to choosing $g = n$, leading to Bayes factors that behave like the BIC and the hyper-g prior of Liang et al. (2005). Foster and George (1994) recommend calibrating the prior based on the risk inflation criterion (RIC) and Fernandez et al. (2001) recommend a combination of the unit information prior and RIC approach. Another alternative is a local empirical Bayes approach, which can be viewed as estimating a separate g for each model, or global empirical Bayes, which assumes a common g but borrows strength from all models (Liang et al., 2005).

For standard deviation parameters in hierarchical models, Gelman (2005) recommends a family of folded-t prior distributions over the commonly used inverse gamma family, due to their flexibility and behavior when random effects are very small. These priors are induced using a parameter-expansion approach which has the added benefit of improving computational efficiency by reducing dependence among the parameters (Liu et al., 1998; Liu and Wu, 1999). This yields a Gibbs sampler less prone to slow mixing when the standard deviations are near zero. The Chen and Dunson (2003) approach had the disadvantages of (1) relying on subjective priors that are difficult to elicit, and (2) computational inefficiency due to slow mixing of the Gibbs sampler; hence we use the parameter-expanded model to address these two problems.

Extending the parameter expansion approach proposed by Gelman (2005) for simple variance component models to the LME model, we replace (3) with

$$y_{ij} = \boldsymbol{X}'_{ij}\boldsymbol{\beta} + \boldsymbol{Z}'_{ij}\boldsymbol{A}\boldsymbol{\Gamma}\boldsymbol{\xi}_i + \epsilon_{ij}, \quad \epsilon_{ij} \sim N(0, \sigma^2), \tag{4}$$

where $\boldsymbol{\xi}_i \sim N(\mathbf{0}, \boldsymbol{D})$ and $\boldsymbol{A} = \mathrm{diag}(\alpha_1, \ldots, \alpha_q)'$ and $\boldsymbol{D} = \mathrm{diag}(d_1, \ldots, d_q)'$ are diagonal matrices, $\alpha_l \sim N(0, 1), l = 1, \ldots, q$, and $d_l \sim IG(\frac{1}{2}, \frac{N}{2})$, $l = 1, \ldots, q$, IG denoting the inverse gamma distribution. Note that the latent random effects have been multiplied by a redundant multiplicative parameter. In this case, the implied covariance decomposition is $\boldsymbol{\Omega} = \boldsymbol{A}\boldsymbol{\Gamma}\boldsymbol{D}\boldsymbol{\Gamma}'\boldsymbol{A}$.

The parameters $\alpha_l,\ l = 1, \ldots, q$, are proportional to λ_l and thus to the random effects standard deviations, so setting $\alpha_l = 0$ effectively drops out the random effects for the lth predictor. When random effects are assumed to be uncorrelated, i.e., $\Gamma = I$ and $\lambda_l, l = 1, \ldots, q$ equal the random effects standard deviations, a folded t prior on $\lambda_l = |\alpha_l|\sqrt{d_l},\ l = 1, \ldots, q$ is induced, as described in Gelman (2005). Generalizing to the case of correlated random effects, a folded-t prior is not induced; however, improved computational efficiency is still achieved, as illustrated in Sect. 6.

In our proposed prior structure, we use a Zellner-type prior for the fixed effects components. Specifically, we let $\boldsymbol{\beta}_J \sim N\left(\mathbf{0}, \sigma^2(\mathbf{X}^{\mathbf{J}\prime}\mathbf{X}^{\mathbf{J}})^{-1}/g\right)$, $g \sim G(\frac{1}{2}, \frac{N}{2})$, $\sigma^2 \propto \frac{1}{\sigma^2}$, and $J_l \sim Be(p_0), l = 1, \ldots, p$, with Be denoting the Bernoulli distribution and $G(a, b)$ denoting the Gamma distribution with mean a/b and variance a/b^2. We give $\alpha_l, l = 1, \ldots, q$, a zero-inflated half-normal prior, $ZI - N^+(0, 1, p_{l0})$, where p_{l0} is the prior probability that $\alpha_l = 0$. Lastly, the free elements of $\boldsymbol{\Gamma}$ are treated as a $q(q-1)/2$-vector with prior $p(\boldsymbol{\gamma}|\boldsymbol{\alpha}) = N(\boldsymbol{\gamma}_0, \boldsymbol{V}_\gamma) \cdot 1(\boldsymbol{\gamma} \in \boldsymbol{R}_{\boldsymbol{\alpha}})$ where $\boldsymbol{R}_{\boldsymbol{\alpha}}$ constrains elements of $\boldsymbol{\gamma}$ to be zero when the corresponding random effects are zero. For simplicity, we do not allow uncertainty in which random effects are correlated.

4.2 Posterior Computation

The joint posterior distribution for $\boldsymbol{\theta} = (\boldsymbol{\alpha}, \boldsymbol{\beta}, \boldsymbol{\gamma}, \sigma^2)$ is given by

$$p(\boldsymbol{\theta}|y) \propto \prod_{i=1}^{n} N_p(\boldsymbol{\xi}_i; \mathbf{0}, \boldsymbol{D}) \prod_{j=1}^{n_i} \left\{N(y_{ij}; \boldsymbol{X}_{ij}'\boldsymbol{\beta} + \boldsymbol{Z}_{ij}'\boldsymbol{A}\boldsymbol{\Gamma}\boldsymbol{\xi}_i, \sigma^2)\right\} p(\sigma^2)p(\boldsymbol{\beta}, \boldsymbol{J}, g)p(\boldsymbol{\alpha}, \boldsymbol{\gamma})p(\boldsymbol{D}). \tag{5}$$

This distribution has a complex form, which we cannot sample from directly; instead, we employ a parameter-expanded Gibbs sampler (Liu et al., 1998; Liu and Wu, 1999). The Gibbs sampler proceeds by iteratively sampling from the full conditional distributions of all parameters $\boldsymbol{\alpha}, \boldsymbol{\gamma}, \boldsymbol{\beta}, \sigma^2$, hyperparameters g and $\boldsymbol{J}$, and the diagonal elements $d_l, l = 1, \ldots, q$ of $\boldsymbol{D}$. The full conditional posterior distributions follow from (5) using straightforward algebraic routes.

Let $\boldsymbol{\psi}$ be the N-vector such that $\psi_{ij} = y_{ij} - \boldsymbol{Z}_{ij}'\boldsymbol{A}\boldsymbol{\Gamma}\boldsymbol{\xi}_i$. The vector of fixed effects coefficients, β, and effectively, $\boldsymbol{X}$, change dimension from iteration to iteration, depending on the value of J, so care needs to be taken to ensure that the dimensions are consistent. Let $\boldsymbol{X}_{ij}^{\mathbf{J}}$ denote the subvector of X_{ij}, $\{X_{ijl} : J_l = 1\}$, $\boldsymbol{\beta}_{\mathbf{J}}$ the subvector β, $\{\beta_l : J_l = 1\}$. The full conditional posterior $p(\boldsymbol{\beta}_J|\boldsymbol{J}, \boldsymbol{\alpha}, \boldsymbol{\gamma}, \sigma^2, \boldsymbol{\xi}, \boldsymbol{y}, \boldsymbol{X}, \boldsymbol{Z})$ is $N(\hat{\boldsymbol{\beta}}_{\mathbf{J}}, \boldsymbol{V}_{\mathbf{J}})$ where

$$\hat{\boldsymbol{\beta}}_{\mathbf{J}} = \left(\sum_{i=1}^{n}\sum_{j=1}^{n_i}(y_{ij} - \boldsymbol{Z}_{ij}'\boldsymbol{A}\boldsymbol{\Gamma}\boldsymbol{\xi}_i)\mathbf{X}^{\mathbf{J}\prime}_{ij}\right) \cdot \frac{\boldsymbol{V}_{\mathbf{J}}}{\sigma^2} \text{ and } \boldsymbol{V}_{\mathbf{J}} = \left(\sum_{i=1}^{n}\sum_{j=1}^{n_i}\mathbf{X}^{\mathbf{J}}_{ij}\mathbf{X}^{\mathbf{J}\prime}_{ij}\left(\frac{1}{\sigma^2} + g\right)\right)^{-1}.$$

To calculate the posterior for $\boldsymbol{J}$ each J_l needs to be updated individually, conditional on J_{-l}, the subvector of J excluding J_l. We calculate $p(J_l = 1|\boldsymbol{J}_{-l}, \boldsymbol{\alpha}, \boldsymbol{\gamma}, \sigma^2, \boldsymbol{\xi}, \boldsymbol{y}, \boldsymbol{X}, \boldsymbol{Z})$ for $l = 1, \ldots, p$, by integrating out $\boldsymbol{\beta}$ and σ^2 as in Smith and Kohn (1996) and obtaining $p(J_l = 1|\boldsymbol{J}_{-l}, \boldsymbol{\alpha}, \boldsymbol{\gamma}, \boldsymbol{\phi}, \boldsymbol{\xi}, \boldsymbol{y}, \boldsymbol{X}, \boldsymbol{Z}) = \frac{1}{1+h_l}$, where $J_{-l} = \{J_i : i \neq l\}$,

$$h_l = \frac{1 - p_{l0}}{p_{l0}} \cdot \left(1 + \frac{1}{g}\right)^{1/2} \cdot \frac{S(J_l = 0)}{S(J_l = 1)}$$

and

$$S(\boldsymbol{J}) = \left(\boldsymbol{\psi}'\boldsymbol{\psi} - \hat{\boldsymbol{\beta}}_{\mathbf{J}} \boldsymbol{V}_{\mathbf{J}}^{-1} \hat{\boldsymbol{\beta}}_{\mathbf{J}}\right)^{-N/2}.$$

$S(J_l = 0)$ is equivalent to $S(\boldsymbol{J})$ but with the element J_l of $\boldsymbol{J}$ set to 0, so $\boldsymbol{\psi}$, $\mathbf{X}^{\mathbf{J}}$, $\hat{\boldsymbol{\beta}}_{\mathbf{J}}$ and $\boldsymbol{V}_{\mathbf{J}}$ may need to be recomputed to correspond to $J_l = 0$. Similarly, for $S(J_l = 1)$.

To complete the fixed effects component updating, the posteriors of g and σ^2 are needed. The gamma and inverse gamma priors used yield conjugate gamma and inverse gamma posteriors. The full conditional posterior for g is given by

$$\Gamma\left(\frac{p_{\mathbf{J}} + 1}{2}, \frac{\boldsymbol{\beta}_J{}'\mathbf{X}^{\mathbf{J}\prime}\mathbf{X}^{\mathbf{J}}\boldsymbol{\beta}_J/\sigma^2 + N}{2}\right)v$$

where $p_{\mathbf{J}} = \sum_{l=1}^{p} 1(J_l = 1)$. The full conditional posterior for σ^2 is given by

$$IG\left(\frac{N + p_{\mathbf{J}}}{2}, \frac{\boldsymbol{\psi}'\boldsymbol{\psi} + g\boldsymbol{\beta}_J{}'\mathbf{X}^{\mathbf{J}\prime}\mathbf{X}^{\mathbf{J}}\boldsymbol{\beta}_J}{2}\right).$$

For the random effects component, the dimensionality does not change between iterations. For γ and ξ_i, the normal priors yield conjugate normal posteriors, while the zero-inflated half-normal prior for each α_l yields a zero-inflated half-normal posterior. Let $\boldsymbol{\psi}$ be the N-vector such that $\psi_{ij} = y_{ij} - \mathbf{X}^{\mathbf{J}}{}_{ij}\boldsymbol{\beta} - \boldsymbol{Z}'_{ij}\boldsymbol{A}\boldsymbol{\Gamma}\boldsymbol{\xi}_i$. The full conditional posterior $p(\boldsymbol{\gamma}|\boldsymbol{\alpha}, \boldsymbol{\beta}, \boldsymbol{\lambda}, \boldsymbol{\xi}, \sigma^2, \boldsymbol{y}, \boldsymbol{X}, \boldsymbol{Z})$ is given by $N(\hat{\boldsymbol{\gamma}}, \hat{\boldsymbol{V}}_\gamma) \cdot 1(\boldsymbol{\gamma} \in \boldsymbol{R}_{\boldsymbol{\lambda}})$ where

$$\hat{\boldsymbol{V}}_\gamma = \left(\sum_{i=1}^{n}\sum_{j=1}^{n_i} \frac{1}{\sigma^2}\boldsymbol{u}_{ij}\boldsymbol{u}'_{ij} + \boldsymbol{V}_{\boldsymbol{\gamma}}^{-1}\right)^{-1} \text{ and } \hat{\boldsymbol{\gamma}} = \left(\sum_{i=1}^{n}\sum_{j=1}^{n_i} \frac{1}{\sigma^2}(y_{ij} - \mathbf{X}^{\mathbf{J}}{}_{ij}\boldsymbol{\beta}_J)\boldsymbol{u}'_{ij} + \boldsymbol{\gamma}_0\boldsymbol{V}_{\boldsymbol{\gamma}}^{-1}\right) \cdot \hat{\boldsymbol{V}}_{\boldsymbol{\gamma}}.$$

The $q(q-1)/2$ vector $\boldsymbol{u}_{ij}$ is defined as $(\xi_{il}\alpha_m Z_{ijm} : l = 1, \ldots, q, m = l + 1, \ldots, q)'$ so that the random effects term $\boldsymbol{Z}'_{ij}\boldsymbol{A}\boldsymbol{\Gamma}\boldsymbol{\xi}_i$ can be written as $\boldsymbol{u}'_{ij}\boldsymbol{\gamma}$.

Each α_l must be updated individually. The zero-inflated truncated normal prior for α_l yields a conjugate posterior $p(\alpha_l|\boldsymbol{\alpha}_{-l}, \boldsymbol{\beta}, \boldsymbol{\gamma}, \boldsymbol{\xi}, \boldsymbol{\phi}, \boldsymbol{y}, \boldsymbol{X}, \boldsymbol{Z}) = ZI - N^{+}(\hat{\alpha}, V_{\alpha l}, \hat{p}_l)$ where

$$\hat{\alpha} = \left(\frac{\sum_{i=1}^{n} \sum_{j=1}^{n_i} t_{ijl} T_{ij}}{\sigma^2} \right) V_{al}, \qquad V_{al} = \left(\sum_{i=1}^{n} \sum_{j=1}^{n_i} \frac{t_{ijl}^2}{\sigma^2} + 1 \right)^{-1},$$

$$\hat{p}_l = \frac{p_{al}}{p_{al} + (1 - p_{al}) \frac{N(0;0,1)}{N(0;\hat{\alpha},V_{al})} \cdot \frac{1-\Phi(0;\hat{\alpha},V_{al})}{1-\Phi(0;0,1)}},$$

where $T_{ij} = y_{ij} - \mathbf{X^J}'_{ij} \boldsymbol{\beta_J} - \sum_{k \neq l} t_{ijk} \alpha_k$ and $N(0; m, v)$ denotes the normal density with mean m and variance v evaluated at 0 and $\Phi(0; m, v)$ is the normal cumulative distribution function with mean m and variance v evaluated at 0. The q vector

$$\boldsymbol{t}_{ij} = \left(Z_{ijl} \left(\xi_{il} + \sum_{m=1}^{l-1} \xi_{im} \gamma_{ml} \right) : l = 1, \ldots, q \right)^T$$

is defined so that the random effects term $\boldsymbol{Z}'_{ij} \boldsymbol{A} \boldsymbol{\Gamma} \boldsymbol{\xi}_i$ can be written as $\boldsymbol{t}'_{ij} \boldsymbol{\alpha}$.

The latent variables $\boldsymbol{\xi}_i, i = 1, \ldots, n$ have posterior $p(\boldsymbol{\xi}_i | \boldsymbol{\beta}, \boldsymbol{\alpha}, \boldsymbol{\gamma}, \sigma^2, \boldsymbol{y}, \boldsymbol{X}, \boldsymbol{Z})$ given by $N(\hat{\boldsymbol{\xi}}_i, \boldsymbol{V}_\xi)$ where

$$\hat{\boldsymbol{\xi}}_i = \sum_{j=i}^{n_i} (y_{ij} - \mathbf{X^J}'_{ij} \boldsymbol{\beta_J}) \boldsymbol{Z}'_{ij} \boldsymbol{A} \boldsymbol{\Gamma} \boldsymbol{V}_\xi \sigma^{-2} \text{ and } \boldsymbol{V}_\xi = \left(\sum_{j=1}^{n_i} \boldsymbol{\Gamma}' \boldsymbol{A} \boldsymbol{Z}_{ij} \boldsymbol{Z}'_{ij} \boldsymbol{A} \boldsymbol{\Gamma} \sigma^{-2} + \boldsymbol{D}^{-1} \right)^{-1}.$$

Only the components of $\boldsymbol{\xi}_i$ corresponding to $\alpha_l > 0$ are updated. Lastly, the diagonal elements of $\boldsymbol{D}$ have inverse gamma priors $IG(\frac{1}{2}, \frac{N}{2})$; hence the posterior is given by

$$p(d_l | \boldsymbol{\alpha}, \boldsymbol{\beta}, \boldsymbol{\gamma}, \boldsymbol{\xi}, \sigma^2, \boldsymbol{y}) = IG\left(\frac{1}{2} + \frac{n}{2}, \frac{N}{2} + \frac{\sum_{i=1}^{n} \xi_{il}^2}{2} \right).$$

The initial MCMC draws, prior to the convergence of the chain, are discarded, and the remaining draws used to obtain posterior summaries of model parameters. Models with high posterior probability can be identified as those appearing most often in the output and considered for further evaluation. Marginal inclusion probabilities for a given coefficient may also be calculated using the proportion of draws in which the coefficient is nonzero.

5 Binary Logistic Mixed Models

Logistic mixed models are widely used, flexible models for unbalanced repeated measures data. Our approach for logistic mixed models is to formulate the model in such a way that its coefficients are conditionally linear and the SSVS approach can again be applied. This entails the use of a data augmentation strategy and approximation of the logistic density, with approximation error corrected for using importance weights. The covariance decomposition in (1) and parameter expansion approach described in Sect. 4.1 are again used.

Defining terms as in (3), the logistic mixed model for a binary response variable y is written as

$$\operatorname{logit}\left(P(y_{ij}=1|\boldsymbol{X}_{ij},\boldsymbol{Z}_{ij},\boldsymbol{\beta},\boldsymbol{a}_i)\right)=\boldsymbol{X}_{ij}'\boldsymbol{\beta}+\boldsymbol{Z}_{ij}'\boldsymbol{a}_i,\quad \boldsymbol{a}_i\sim N(\boldsymbol{0},\boldsymbol{\Omega}). \tag{6}$$

We would like to be able to apply the SSVS approach as in the normal case. If we apply the covariance decomposition in (1) to the logistic mixed model, we have

$$\operatorname{logit}\left(P(y_{ij}=1|\boldsymbol{X}_{ij},\boldsymbol{Z}_{ij},\boldsymbol{\beta},\boldsymbol{\lambda},\boldsymbol{\gamma},\boldsymbol{b}_i)\right)=\boldsymbol{X}_{ij}'\boldsymbol{\beta}+\boldsymbol{Z}_{ij}'\boldsymbol{\Lambda}\boldsymbol{\Gamma}\boldsymbol{b}_i,\quad \boldsymbol{b}_i\sim N(\boldsymbol{0},\boldsymbol{I}). \tag{7}$$

In this case, the model is nonlinear and we do not immediately have conditional linearity for the variance parameters $\boldsymbol{\lambda}$ and $\boldsymbol{\gamma}$ as in the normal case. To obtain conditional linearity for the model coefficients, we take advantage of the fact that the logistic distribution can be closely approximated by the t distribution (Albert and Chib, 1993; Holmes and Knorr-Held, 2003; O'Brien and Dunson, 2004), and that the t distribution can be expressed as a scale mixture of normals (West, 1987).

First, note that (7) is equivalent to the specification

$$y_{ij}=\begin{cases}1 & w_{ij}>0\\ 0 & w_{ij}\le 0\end{cases},$$

where w_{ij} is a logistically distributed random variable with location parameter $\boldsymbol{X}_{ij}'\boldsymbol{\beta}+\boldsymbol{Z}_{ij}'\boldsymbol{\Lambda}\boldsymbol{\Gamma}\boldsymbol{b}_i$ and density function

$$\mathcal{L}(w_{ij}|\boldsymbol{X}_{ij},\boldsymbol{Z}_{ij},\boldsymbol{\beta},\boldsymbol{\lambda},\boldsymbol{\gamma})=\frac{\exp\{-(w_{ij}-\boldsymbol{X}_{ij}'\boldsymbol{\beta}-\boldsymbol{Z}_{ij}'\boldsymbol{\Lambda}\boldsymbol{\Gamma}\boldsymbol{b}_i)\}}{\{1+\exp[-(w_{ij}-\boldsymbol{X}_{ij}'\boldsymbol{\beta}-\boldsymbol{Z}_{ij}'\boldsymbol{\Lambda}\boldsymbol{\Gamma}\boldsymbol{b}_i)]\}^2}.$$

Then, as w_{ij} is approximately distributed as a noncentral t_ν with location parameter $\boldsymbol{X}_{ij}'\boldsymbol{\beta}+\boldsymbol{Z}_{ij}'\boldsymbol{\Lambda}\boldsymbol{\Gamma}\boldsymbol{b}_i$ and scale parameter $\tilde{\sigma}^2$, we can express it as a scale mixture of normals and write

$$w_{ij}=\boldsymbol{X}_{ij}'\boldsymbol{\beta}+\boldsymbol{Z}_{ij}'\boldsymbol{\Lambda}\boldsymbol{\Gamma}\boldsymbol{b}_i+\epsilon_{ij},\quad \epsilon_{ij}\sim N(0,\tilde{\sigma}^2/\phi_{ij}), \tag{8}$$

where $\phi_{ij}\sim G\left(\frac{\nu}{2},\frac{\nu}{2}\right)$. Setting $\nu=7.3$ and $\tilde{\sigma}^2=\pi^2(\nu-2)/3\nu$ makes the approximation nearly exact. The approximation error, though negligible except in the extreme tails, may be corrected for by importance weighting when making inferences. Under this model formulation, we have a model in which all coefficients are conditionally normal, and we are able to apply SSVS to the problem. We also are able to take advantage of the improved computational efficiency of a parameter expanded model as in (4). Applying the parameter expansion to (8) we have

$$w_{ij}=\boldsymbol{X}_{ij}'\boldsymbol{\beta}+\boldsymbol{Z}_{ij}\boldsymbol{A}\boldsymbol{\Gamma}\boldsymbol{\xi}_i+\epsilon_{ij},\quad \epsilon_{ij}\sim N(0,\tilde{\sigma}^2/\phi_{ij}),$$

where terms are defined as in (4) and (8). We will use this model formulation to propose a prior structure and compute posterior distributions.

5.1 Priors and Posterior Computation

We use the same priors for the random effects parameters as in the normal case, and similar priors for the fixed effects parameters. We specify $\boldsymbol{\beta}_J \sim N\left(0, (\mathbf{X^{J'}X^J})^{-1}/g\right)$, $g \sim G(\frac{1}{2}, \frac{N}{2})$, and $J_l \sim Be(p_0), l = 1, \ldots, p$. Using the t-distribution to approximate the likelihood as previously described, the joint posterior distribution for $\theta = (\boldsymbol{\alpha}, \boldsymbol{\beta}, \boldsymbol{\gamma}, \boldsymbol{\phi})$ is given by

$$p(\boldsymbol{\theta}|\mathbf{y}) \propto p(\boldsymbol{\beta}, \boldsymbol{J}, g)p(\boldsymbol{\gamma}, \boldsymbol{\alpha})p(\boldsymbol{D})\left(\prod_{i=1}^{n} N_q(\boldsymbol{\xi}_i; \mathbf{0}, \boldsymbol{D}) \prod_{j=1}^{n_i} \left[N\left(w_{ij}; \boldsymbol{X}_{ij}\boldsymbol{\beta} + \mathbf{Z}_{ij}\boldsymbol{A\Gamma\xi}_i, \frac{\tilde{\sigma}^2}{\phi_{ij}}\right) \times\{1(w_{ij} > 0)y_{ij} + 1(w_{ij} \le 0) (1 - y_{ij})\}p(\phi_{ij})\right]\right). \quad (9)$$

Again we have a complex posterior from which we cannot directly sample and we employ a Gibbs sampler. In introducing a latent variable w_{ij} we have applied a data augmentation strategy related to Albert and Chib (1993) and used for multivariate logistic models by O'Brien and Dunson (2004). This auxiliary variable is updated in the Gibbs sampler and its full conditional posterior follows immediately from (9) as a normal distribution truncated above or below by 0 depending on y_{ij}

$$p(w_{ij}|\boldsymbol{\theta}, y_{ij}) = \frac{N\left(w_{ij}; \boldsymbol{X}_{ij}\boldsymbol{\beta} + \mathbf{Z}'_{ij}\boldsymbol{A\Gamma\xi}_i, \frac{\tilde{\sigma}^2}{\phi_{ij}}\right) \cdot 1\left((-1)^{y_{ij}} w_{ij} < 0\right)}{\Phi\left(0; \boldsymbol{X}'_{ij}\boldsymbol{\beta} + \mathbf{Z}'_{ij}\boldsymbol{A\Gamma\xi}_i, \frac{\tilde{\sigma}^2}{\phi_{ij}}\right)^{1-y_{ij}} \left\{1 - \Phi\left(0; \boldsymbol{X}'_{ij}\boldsymbol{\beta} + \mathbf{Z}'_{ij}\boldsymbol{A\Gamma\xi}_i, \frac{\tilde{\sigma}^2}{\phi_{ij}}\right)\right\}^{y_{ij}}}, \quad (10)$$

where $\Phi(\cdot)$ indicates the normal cumulative distribution function.

The Gibbs sampler proceeds by iteratively sampling from the full conditional distributions of w and all parameters $\boldsymbol{\alpha}, \boldsymbol{\gamma}, \boldsymbol{\beta}, \boldsymbol{\phi}$, hyperparameters g and $\boldsymbol{J}$, as well as the latent variable $\boldsymbol{\xi}_i, i = 1, \ldots, n$ and the diagonal elements d_l, $l = 1, \ldots, q$ of $\boldsymbol{D}$. The remaining full conditional posterior distributions follow from (9) and are similar in form to the normal case. Some differences are that σ^2 is fixed and the Gibbs sampler additionally updates w and ϕ.

Let $\boldsymbol{\psi}$ be the N-vector such that $\psi_{ij} = w_{ij} - \mathbf{Z}'_{ij}\boldsymbol{A\Gamma\xi}_i$. As in the normal case, the vector of fixed effects coefficients, β, and effectively, $\boldsymbol{X}$, change dimension from iteration to iteration, depending on the value of $\boldsymbol{J}$, so care needs to be taken to ensure that the dimensions are consistent. Let $\boldsymbol{X}^{\mathbf{J}}_{ij}$ denote the subvector of X_{ij}, $\{X_{ijl} : J_l = 1\}$, $\boldsymbol{\beta}_{\mathbf{J}}$ the subvector β, $\{\beta_l : J_l = 1\}$. The full conditional posterior $p(\boldsymbol{\beta}_J|\boldsymbol{J}, \boldsymbol{\alpha}, \boldsymbol{\gamma}, \boldsymbol{\phi}, \boldsymbol{\xi}, y, X, Z)$ is $N(\hat{\boldsymbol{\beta}}_{\mathbf{J}}, \boldsymbol{V}_{\mathbf{J}})$ where

$$\hat{\boldsymbol{\beta}}_{\mathbf{J}} = \left(\sum_{i=1}^{n}\sum_{j=1}^{n_i} \frac{\phi_{ij}}{\tilde{\sigma}^2}\psi_{ij}\mathbf{X^{J'}}_{ij}\right) \cdot \boldsymbol{V}_{\mathbf{J}} \text{ and } \boldsymbol{V}_{\mathbf{J}} = \left(\sum_{i=1}^{n}\sum_{j=1}^{n_i} \mathbf{X^J}_{ij}\mathbf{X^{J'}}_{ij}\left(\frac{\phi_{ij}}{\tilde{\sigma}^2} + g\right)\right)^{-1}.$$

To calculate the posterior for $\boldsymbol{J}$ each J_l needs to be updated individually. We calculate $p(J_l = 1|\boldsymbol{J}_{-l}, \boldsymbol{\alpha}, \boldsymbol{\gamma}, \boldsymbol{\phi}, \boldsymbol{\xi}, \boldsymbol{y}, \boldsymbol{X}, \boldsymbol{Z})$ for $l = 1, \ldots, p$, by integrating out $\boldsymbol{\beta}$ as in Smith and Kohn (1996) and obtaining $p(J_l = 1|\boldsymbol{J}_{-l}, \boldsymbol{\alpha}, \boldsymbol{\gamma}, \boldsymbol{\phi}, \boldsymbol{\xi}, \boldsymbol{y}, \boldsymbol{X}, \boldsymbol{Z}) = \frac{1}{1+h_l}$, where $J_{-l} = \{J_i : i \neq l\}$,

$$h_l = \frac{1 - p_{l0}}{p_{l0}} \cdot \left(\frac{1}{g}\right)^{1/2} \cdot \frac{S(J_l = 0)}{S(J_l = 1)}$$

and

$$S(\boldsymbol{J}) = |\mathbf{X}^{\mathbf{J}\prime}\mathbf{X}^{\mathbf{J}}|^{1/2} \cdot |\boldsymbol{V}_{\mathbf{J}}|^{1/2} \exp\left\{-\frac{1}{2}\left(\sum_{i=1}^{n}\sum_{j=1}^{n_i} \phi_{ij}\psi_{ij}^2 - \hat{\boldsymbol{\beta}}_{\mathbf{J}}' \boldsymbol{V}_{\mathbf{J}}^{-1} \hat{\boldsymbol{\beta}}_{\mathbf{J}}\right)\right\}.$$

$S(J_l = 0)$ is equivalent to $S(\boldsymbol{J})$ but with the element J_l of $\boldsymbol{J}$ set to 0, so $\boldsymbol{\psi}$, $\mathbf{X}^{\mathbf{J}}$, $\hat{\boldsymbol{\beta}}_{\mathbf{J}}$ and $\boldsymbol{V}_{\mathbf{J}}$ may need to be recomputed to correspond to $J_l = 0$. Similarly for $S(J_l = 1)$.

To complete the fixed effects component updating, the posteriors of g and ϕ are needed. The gamma priors used yield conjugate gamma posteriors. The full conditional posterior for g is given by

$$\Gamma\left(\frac{p_{\mathbf{J}} + 1}{2}, \frac{\boldsymbol{\beta}_J{}'\mathbf{X}^{\mathbf{J}\prime}\mathbf{X}^{\mathbf{J}}\boldsymbol{\beta}_J + N}{2}\right),$$

where $p_{\mathbf{J}} = \sum_{l=1}^{p} 1(J_l = 1)$. The components of ϕ, ϕ_{ij}, are not identically distributed. Each ϕ_{ij} has a conjugate gamma posterior

$$G\left(\frac{\nu + 1}{2}, \frac{(w_{ij} - \boldsymbol{Z}_{ij}\boldsymbol{A}\boldsymbol{\Gamma}\boldsymbol{\xi}_i - \boldsymbol{X}_{ij}'\boldsymbol{\beta})^2/\tilde{\sigma}^2 + \nu}{2}\right).$$

For the random effects component, the dimensionality does not change between iterations. For γ and ξ_i, the normal priors yield conjugate normal posteriors, while the zero-inflated half-normal prior for each α_l yields a zero-inflated half-normal posterior. Let $\boldsymbol{\psi}$ be the N-vector such that $\psi_{ij} = y_{ij} - \mathbf{X^J}_{ij}\boldsymbol{\beta} - \boldsymbol{Z}_{ij}'\boldsymbol{A}\boldsymbol{\Gamma}\boldsymbol{\xi}_i$.

The full conditional posterior $p(\boldsymbol{\gamma}|\boldsymbol{\alpha}, \boldsymbol{\beta}, \boldsymbol{\lambda}, \boldsymbol{\xi}, \boldsymbol{\phi}, \boldsymbol{y}, \boldsymbol{X}, \boldsymbol{Z})$ is given by $N(\hat{\boldsymbol{\gamma}}, \hat{\boldsymbol{V}}_\gamma) \cdot 1(\boldsymbol{\gamma} \in \boldsymbol{R}_{\boldsymbol{\lambda}})$ where

$$\hat{\boldsymbol{V}}_\gamma = \left(\sum_{i=1}^{n}\sum_{j=1}^{n_i} \frac{\phi_{ij}}{\tilde{\sigma}^2} \boldsymbol{u}_{ij}\boldsymbol{u}_{ij}' + \boldsymbol{V}_{\boldsymbol{\gamma}}^{-1}\right)^{-1} \text{ and } \hat{\boldsymbol{\gamma}} = \left(\sum_{i=1}^{n}\sum_{j=1}^{n_i} \frac{\phi_{ij}}{\tilde{\sigma}^2}(w_{ij} - \mathbf{X^J}_{ij}\boldsymbol{\beta}_J)\boldsymbol{u}_{ij}' + \boldsymbol{\gamma}_0\boldsymbol{V}_{\boldsymbol{\gamma}}^{-1}\right) \cdot \hat{\boldsymbol{V}}_{\boldsymbol{\gamma}}.$$

The $q(q-1)/2$ vector $\boldsymbol{u}_{ij}$ is defined as $(\xi_{il}\alpha_m Z_{ijm} : l = 1, \ldots, q, m = l + 1, \ldots, q)'$ so that the random effects term $\boldsymbol{Z}_{ij}'\boldsymbol{A}\boldsymbol{\Gamma}\boldsymbol{\xi}_i$ can be written as $\boldsymbol{u}_{ij}'\boldsymbol{\gamma}$.

Each α_l must be updated individually. The zero-inflated truncated normal prior for α_l yields a conjugate posterior $p(\alpha_l|\boldsymbol{\alpha}_{-l}, \boldsymbol{\beta}, \boldsymbol{\gamma}, \boldsymbol{\xi}, \boldsymbol{\phi}, \boldsymbol{y}, \boldsymbol{X}, \boldsymbol{Z}) = ZI - N^{+}(\hat{\alpha}, V_{\alpha l}, \hat{p}_l)$ where

$$\hat{\alpha} = \left(\frac{\sum_{i=1}^{n} \sum_{j=1}^{n_i} \phi_{ij} t_{ijl} T_{ij}}{\tilde{\sigma}^2} \right) V_{al}, \qquad V_{al} = \left(\sum_{i=1}^{n} \sum_{j=1}^{n_i} \frac{\phi_{ij} t_{ijl}^2}{\tilde{\sigma}^2} + 1 \right)^{-1},$$

$$\hat{p}_l = \frac{p_{al}}{p_{al} + (1 - p_{al}) \frac{N(0;0,1)}{N(0;\hat{\alpha},V_{al})} \cdot \frac{1-\Phi(0;\hat{\alpha},V_{al})}{1-\Phi(0;0,1)}},$$

where $T_{ij} = w_{ij} - \mathbf{X^{J}}'_{ij}\boldsymbol{\beta}_J - \sum_{k \neq l} t_{ijk}\alpha_k$ and $N(0; m, v)$ denotes the normal density with mean m and variance v evaluated at 0 and $\Phi(0; m, v)$ is the normal cumulative distribution function with mean m and variance v evaluated at 0. The q vector

$$\boldsymbol{t}_{ij} = \left(Z_{ijl} \left(\xi_{il} + \sum_{m=1}^{l-1} \xi_{im} \gamma_{ml} \right) : l = 1, \ldots, q \right)^T$$

is defined so that the random effects term $\mathbf{Z}'_{ij}\boldsymbol{A\Gamma\xi}_i$ can be written as $\boldsymbol{t}'_{ij}\boldsymbol{\alpha}$.

The latent variables $\boldsymbol{\xi}_i$, $i = 1, \ldots, n$ have posterior $p(\boldsymbol{\xi}_i | \boldsymbol{\beta}, \boldsymbol{\alpha}, \boldsymbol{\gamma}, \boldsymbol{\phi}, \boldsymbol{y}, \boldsymbol{X}, \boldsymbol{Z})$ given by $N(\hat{\boldsymbol{\xi}}_i, \boldsymbol{V}_\xi)$ where

$$\hat{\xi}_i = \sum_{j=i}^{n_i} \phi_{ij}(w_{ij} - \mathbf{X^{J}}'_{ij}\boldsymbol{\beta}_J) \boldsymbol{Z}'_{ij}\boldsymbol{A\Gamma V}_\xi \tilde{\sigma}^{-2} \text{ and } \boldsymbol{V}_\xi = \left(\sum_{j=1}^{n_i} \phi_{ij} \boldsymbol{\Gamma}'\boldsymbol{A Z}_{ij}\boldsymbol{Z}'_{ij}\boldsymbol{A\Gamma}\tilde{\sigma}^{-2} + \boldsymbol{D}^{-1} \right)^{-1}.$$

Only the components of $\boldsymbol{\xi}_i$ corresponding to $\alpha_l > 0$ are updated. Lastly, the diagonal elements of $\boldsymbol{D}$ have inverse gamma priors $IG(\frac{1}{2}, \frac{N}{2})$; hence the posterior is given by $p(d_l | \boldsymbol{\alpha}, \boldsymbol{\beta}, \boldsymbol{\gamma}, \boldsymbol{\xi}, \boldsymbol{\phi}, \boldsymbol{y}) = IG\left(\frac{1}{2} + \frac{n}{2}, \frac{N}{2} + \frac{\sum_{i=1}^{n} \xi_{il}^2}{2} \right)$.

5.2 Importance Weights

This Gibbs sampler generates samples from an approximate posterior as we have approximated the logistic likelihood in (8). To correct for this, importance weights (Hastings, 1970) may be applied when computing posterior summaries to obtain exact inferences. If we have M iterations of our Gibbs sampler, excluding the burn-in interval, then our importance weights $r^{(t)}$, $t = 1, \ldots, M$ can be computed as

$$r^{(t)} = \prod_{i=1}^{n} \prod_{j=1}^{n_i} \frac{\mathcal{L}(w_{ij}; \boldsymbol{X}'_{ij}\boldsymbol{\beta} + \boldsymbol{Z}'_{ij}\boldsymbol{A\Gamma\xi}_i)}{\mathcal{T}_\nu(w_{ij}; \boldsymbol{X}'_{ij}\boldsymbol{\beta} + \boldsymbol{Z}'_{ij}\boldsymbol{A\Gamma\xi}_i, \tilde{\sigma}^2)},$$

where $\mathcal{L}(\cdot)$ is the logistic density function and $\mathcal{T}_\nu(\cdot)$ is the t density function with degrees of freedom ν.

Posterior means, probabilities, and other summaries of the model parameters can be estimated from the Gibbs sampler output using an importance-weighted sample average. For example, the posterior probability for a given model m is the sum of the weights corresponding to each occurrence of model m in the posterior sample, divided by the sum of all M weights. The approximation is very close and hence the weights are close to one. In our simulation and data examples, we found very little difference between weighted and unweighted results.

In lieu of approximating the logistic distribution with the t distribution, we also considered the slice sampler for sampling from the exact posterior distribution as applied by Gerlach et al. (2002) to variable selection for logistic models. In this approach, the model is considered linear with response variable $v_{ij} = \text{logit}\left(p(y_{ij}=1)\right)$, the vector of log odds, and $v_{ij} = \text{logit}\left(p(y_{ij}=1)\right) = \boldsymbol{X}'_{ij}\boldsymbol{\beta} + \boldsymbol{Z}'_{ij}\Lambda\Gamma\boldsymbol{b}_i + \epsilon_{ij}, \epsilon_{ij} \sim N(0,\sigma^2)$. The vector v_{ij} is updated in a data-augmented Gibbs sampler where an auxiliary variable $u_{ij} \sim U\left(0, \frac{1}{1+\exp(v_{ij})}\right)$ is introduced so that the full conditional posterior distribution for v_{ij} is simplified to a truncated normal distribution as follows:

$$
\begin{aligned}
p(v_{ij}|y_{ij},\boldsymbol{X}_{ij},\boldsymbol{Z}_{ij},\boldsymbol{\beta},\boldsymbol{\alpha},\boldsymbol{\gamma},\sigma^2) &\propto p(y_{ij}|v_{ij})\cdot p(v_{ij}|\boldsymbol{X}_{ij},\boldsymbol{Z}_{ij},\boldsymbol{\beta},\boldsymbol{\alpha},\boldsymbol{\gamma},\sigma^2)\\
&\propto \left(\frac{e^{v_{ij}y_{ij}}}{1+e^{v_{ij}}}\right)\cdot N(\boldsymbol{X}'_{ij}\boldsymbol{\beta}+\boldsymbol{Z}'_{ij}\boldsymbol{A}\Gamma\boldsymbol{\xi}_i,\sigma^2),\\
p(v_{ij}|u_{ij},\boldsymbol{X}_{ij},\boldsymbol{Z}_{ij},\boldsymbol{\beta},\boldsymbol{\alpha},\boldsymbol{\gamma},\sigma^2) &\propto p(u_{ij}|v_{ij})p(v_{ij}|\boldsymbol{X}_{ij},\boldsymbol{Z}_{ij},\boldsymbol{\beta},\boldsymbol{\alpha},\boldsymbol{\gamma},\sigma^2)\\
&\propto N(\boldsymbol{X}'_{ij}\boldsymbol{\beta}+\boldsymbol{Z}'_{ij}\boldsymbol{A}\Gamma\boldsymbol{\xi}_i+\sigma^2 y_{ij},\sigma^2)\cdot 1\left(v_{ij}<\log\left(\frac{1-u_{ij}}{u_{ij}}\right)\right).
\end{aligned}
$$

While slice sampling in general has been noted to have appealing theoretical properties (Neal, 2000; Mira and Tierney, 2002)), it demonstrated unsatisfactory convergence properties for our purposes due to asymmetries induced by the likelihood (Green, 1997). In simulations using the slice sampler approach, the correct models were quickly located; however, the Gibbs chains for nonzero model coefficients were extremely slow to converge.

6 Simulation Examples

We evaluate the proposed approach using a simulation example for a binary response logistic model. We generate three covariates from $U(-2,2)$ for 30 observations on each of 200 subjects, so $\boldsymbol{X}_{ij} = (1, X_{ij1}, X_{ij2}, X_{ij3})'$ and we let $\boldsymbol{Z}_{ij} = \boldsymbol{X}_{ij}$. We let $\boldsymbol{\beta} = (1,0,1,1)'$ and $\alpha_i \sim N(\boldsymbol{0},\boldsymbol{\Omega})$, choosing a range of realistic values for the random effects variances

$$
\boldsymbol{\Omega} = \begin{pmatrix} 0.90 & 0.48 & 0.06 & 0\\ 0.48 & 0.40 & 0.10 & 0\\ 0.06 & 0.10 & 0.10 & 0\\ 0 & 0 & 0 & 0 \end{pmatrix}
$$

We generate $\text{logit}\left(P(y_{ij}=1)\right)$ according to the model (6) and then y_{ij} from $\text{Be}\left(p(y_{ij})\right)$. We follow the prior specification outlined in Sect. 4.1 and induce heavy-tailed priors on the fixed effects coefficients and random effects variances. These default priors do not require subjective choice of hyperparameter values, with the exception of the prior inclusion probabilities, which can be chosen as $p = 0.5$ to give equal probability to inclusion and exclusion, and the prior mean and variance of $\boldsymbol{\gamma}$. Our prior specification does include an informative normal prior for $\boldsymbol{\gamma}$; however, $\boldsymbol{\gamma}$ is scaled in the parameter-expanded model and hence an informative prior can reasonably be chosen. A prior that modestly shrinks the correlations toward zero is desirable for stable estimation while still allowing the data to inform the relationships between the random effects. As a reasonable choice, we set the prior mean and variance for $\boldsymbol{\gamma}$ to be $\mathbf{0}$ and $0.5\boldsymbol{I}$, which can be used as a default in other applications.

We ran the Gibbs sampler for 20,000 iterations, after a burning period of 5,000 iterations. Three chains with dispersed starting values were run and found to converge after a few thousand iterations. The resulting MCMC chains for the random effects variances are shown in Fig. 1 and the posterior means for the fixed effects coefficients and random effects variances are given in Table 1, along with the PQL estimates computed by glmmPQL in R.

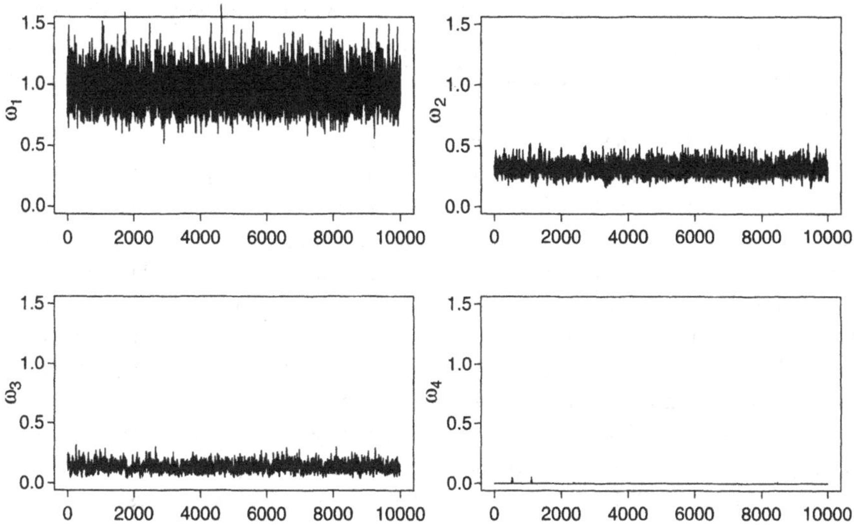

Fig. 1 Gibbs chains for random effects variances

Table 1 Simulation results

Parameter	True value	PQL	95% CI	Post. mean	95% CI	Pr (inclusion)
β_0	1.0	0.901	(0.753, 1.048)	0.892	(0.759, 1.027)	1.000
β_1	0.0	0.031	(−0.062, 0.125)	0.001	(0.000, 0.006)	0.044
β_2	1.0	0.900	(0.820, 0.980)	0.929	(0.845, 1.016)	1.000
β_3	1.0	0.961	(0.896, 1.025)	0.990	(0.920, 1.061)	1.000
ω_1	0.9	0.899		0.958	(0.721, 1.252)	1.000
ω_2	0.4	0.298		0.315	(0.221, 0.427)	1.000
ω_3	0.1	0.143		0.136	(0.072, 0.215)	1.000
ω_4	0.0	0.026		0.000	(0.000, 0.000)	0.008

We compare our results to the penalized quasilikelihood (PQL) approach (Breslow and Clayton, 1993), as this approach is widely used for estimating GLMMs. Although our focus is on selection and inferences in the variance components allowing for model uncertainty, which is not addressed by current frequentist methods, we also obtain model-averaged coefficient estimates. Based on the limited number of simulations run, our estimates tend to be less biased, or closer to the true values than the PQL estimates, which are also known to be biased (Breslow, 2003; Jang and Lim, 2005)). Our algorithm is too computationally intense to run a large enough simulation to definitively assess the frequentist operating characteristics of our approach.

We also compute credible intervals for the random effects variances. To our knowledge, methods for estimating valid frequentist confidence intervals for variance components remain to be developed. In addition we are able to simultaneously compute marginal posterior inclusion probabilities for both the fixed effects and random effects and correctly locate the true model as the one with highest posterior probability.

To evaluate sensitivity to the prior inclusion probability, we also repeated the simulation with prior probabilities set to 0.2 and 0.8 and found very little difference in the posterior means shown in Table 1. Posterior model probabilities were slightly different when the prior inclusion probabilities were changed; however there was no difference in parameter estimates, inferences or model ranking. In each case the true model had the highest posterior probability.

To evaluate the effect of using the priors induced by the parameter expanded model, we compare simulation results between two Gibbs samplers, one including and one excluding the redundant multiplicative parameter in the random effects component. As expected, we do not see any real difference in the point estimates; however, as seen in Fig. 2, the parameter expansion approach resulted in improved computational efficiency and MCMC chains for the random effects variances. Table 2 shows the reduction in autocorrelation in the Gibbs chains. Note we have not directly drawn from the posterior distribution of the variances, rather we have computed them from the MCMC draws for $\boldsymbol{\alpha}$, $\boldsymbol{\gamma}$, $\boldsymbol{\lambda}$ and $\mathbf{d}$. The overparameterization causes the Gibbs chains for these parameters to mix poorly, but in combination they produce well-behaved chains for the random effects variances.

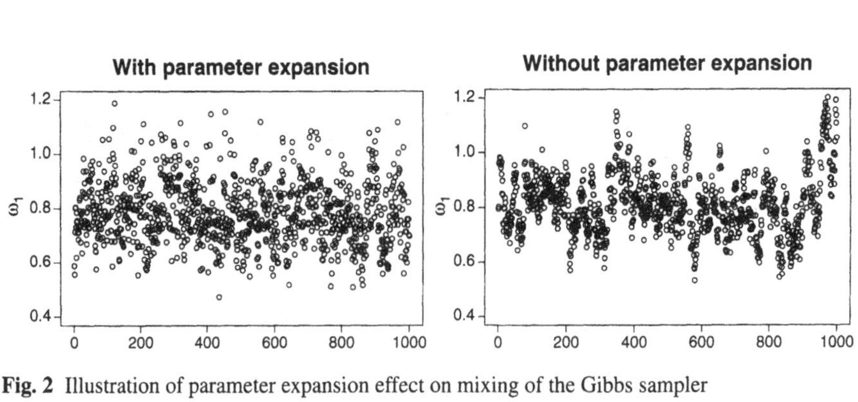

Fig. 2 Illustration of parameter expansion effect on mixing of the Gibbs sampler

Table 2 Autocorrelations in Gibbs chains, with and without parameter expansion

	Lag:	1	2	3	4	5	6	7	8	9	10
ω_1	w/o parameter exp	0.902	0.810	0.726	0.645	0.574	0.511	0.451	0.392	0.342	0.299
	w/parameter exp	0.422	0.350	0.288	0.252	0.208	0.177	0.154	0.142	0.132	0.117
ω_2	w/o parameter exp	0.783	0.653	0.558	0.484	0.422	0.369	0.324	0.286	0.251	0.225
	w/parameter exp	0.563	0.461	0.375	0.326	0.290	0.251	0.222	0.184	0.160	0.148
ω_3	w/o parameter exp	0.853	0.756	0.682	0.618	0.572	0.529	0.487	0.450	0.422	0.393
	w/parameter exp	0.811	0.711	0.639	0.574	0.520	0.477	0.441	0.417	0.388	0.362
ω_4	w/o parameter exp	0.808	0.629	0.439	0.335	0.228	0.162	0.087	0.038	0.008	−0.003
	w/parameter exp	0.595	0.399	0.358	0.295	0.198	−0.001	−0.001	−0.001	−0.001	−0.001

7 Epidemiology Application

As a motivating example, we consider data from the Collaborative Perinatal Project (CPP) conducted between 1959 and 1966. We examine the effect of DDE, a metabolite of DDT, as measured in maternal serum, on pregnancy loss, a binary response variable. Potential confounding variables include mother's age, body mass index, smoking status, and serum levels of cholesterol and triglycerides. Data were collected across 12 different study centers and there is potential for heterogeneity across centers. We are interested in selecting a logistic mixed effects model relating DDE levels and pregnancy loss, accounting for heterogeneity among study centers in those factors that vary in their effects across centers. In addition, inferences on whether predictors such as DDE vary in their effect is of substantial interest.

We let our binary response variable $y_{ij} = 1$ indicate pregnancy loss for participant j in study center i, $i = 1, \ldots, 12$; $j = 1, \ldots, n_i$, for 5,389 total participants. Our covariate vector is $\boldsymbol{X}_{ij} = (1, X_{ij1}, \ldots, X_{ij5})'$ where X_{ij1} is the level of DDE, and $X_{ij2}, \ldots, X_{ij5}$ are the potential confounding variables. All covariates are continuous and centered at their means, and we let $\boldsymbol{Z}_{ij} = \boldsymbol{X}_{ij}$, thus considering all coefficients, including the intercept, for possible heterogeneity among centers.

Priors were chosen as in the simulation example and the Gibbs sampler run for 30,000 iterations after a burning period of 5,000. The Gibbs sampling results indicate that there is no heterogeneity present among study centers and that a fixed effects model is appropriate. The preferred model, as shown in Table 3, includes only the intercept, body mass index, and age, as predictors. The posterior means for all variances are close to zero. A few models with nonzero posterior probability do contain a random effect. The posterior means for the fixed effect are similar to the PQL results returned by glmmPQL in R for the full model, shown in Table 4. These results also show that DDE did not have an appreciable effect on pregancy loss in the CPP study. The PQL results indicate that DDE had a very small but statistically significant effect; however, this may be due to bias in the PQL approach. Applying the BIC criteria to select the best fixed effects model, we obtain the high posterior probability model shown in Table 3.

Table 3 Models with highest posterior probability

Prob	Model
0.58	$X_0, X_{\text{bmi}}, X_{\text{age}}$
0.16	X_0, X_{age}
0.09	$X_0, X_{\text{bmi}}, X_{\text{age}}, X_{\text{dde}}$
0.05	$X_0, X_{\text{chol}}, X_{\text{bmi}}, X_{\text{age}}$
0.03	$X_0, X_{\text{age}}, X_{\text{dde}}$
0.02	$X_0, X_{\text{tg}}, X_{\text{bmi}}, X_{\text{age}}$
0.01	$X_0, X_{\text{bmi}}, X_{\text{age}}, Z_{\text{chol}}$
0.01	$X_0, X_{\text{chol}}, X_{\text{age}}$
0.01	$X_0, X_{\text{chol}}, X_{\text{bmi}}, X_{\text{age}}, X_{\text{dde}}$
0.01	$X_0, X_{\text{age}}, Z_{\text{bmi}}$

Table 4 Posterior summary of fixed effects in CPP example

	PQL	95% CI	Mean	95% CI	$p(\beta_l = 0)$
β_0	−1.813	(−1.943, −1.700)	−1.793	(−1.871, −1.716)	0.000
β_{tg}	0.014	(−0.087, 0.101)	0.000	(0.000, 0.000)	0.968
β_{chol}	−0.081	(−0.219, −0.001)	−0.002	(−0.034, 0.000)	0.932
β_{bmi}	−0.138	(−0.229, −0.055)	−0.096	(−0.210, 0.000)	0.239
β_{age}	0.295	(0.211, 0.372)	0.279	(0.205, 0.352)	0.000
β_{dde}	0.088	(0.009, 0.189)	0.005	(0.000, 0.067)	0.876

8 Other Models

8.1 Logistic Models for Ordinal Data

This framework can also be adapted to accommodate logistic mixed models with ordinal response variables $y_{ij} \in \{1, \ldots, C\}$

$$\text{logit}\left(P(y_{ij} \le c | \boldsymbol{X}_{ij}, \boldsymbol{Z}_{ij}, \boldsymbol{\beta}, \boldsymbol{a}_i, \boldsymbol{\tau})\right) = \tau_c - \boldsymbol{X}'_{ij}\boldsymbol{\beta} - \boldsymbol{Z}'_{ij}\boldsymbol{a}_i, \quad c \in \{1, \ldots, C\}, \tag{11}$$

where terms in the linear predictor are as defined in (3) and $\boldsymbol{\tau} = (\tau_1, \ldots, \tau_{C-1})'$ where $\tau_1 = 0$ for identifiability and $-\infty = \tau_0 < \tau_1 < \cdots < \tau_C = \infty$ are threshold parameters for the ordered categories. Our data augmentation stochastic search Gibbs sampler can be applied to (11) with modifications to truncate w_{ij} to $[\tau_{c-1}, \tau_c]$ for $y_{ij} = c$ and to update the threshold parameters $\boldsymbol{\tau}$. Although updating of $\boldsymbol{\tau}$ can potentially proceed after augmentation as described in Albert and Chib (1993), such an approach has a tendency to mix very slowly (Johnson and Albert, 1999). A modification in which the latent variables $\{w_{ij}\}$ are integrated out and a Metropolis-Hastings step is used, yields better results. An alternative, which allows the baseline parameters $\boldsymbol{\tau}$ to be updated jointly from a multivariate normal posterior after augmentation, is to consider a continuation-ratio logit formulation of the form $\text{logit}\left(P(y_{ij} = c | y_{ij} \ge c, \boldsymbol{X}_{ij}, \boldsymbol{Z}_{ij}, \boldsymbol{\beta}, \boldsymbol{a}_i)\right) = \boldsymbol{X}'_{ij}\boldsymbol{\beta} + \boldsymbol{Z}'_{ij}\boldsymbol{a}_i$, instead of (11) (Agresti, 1990). Such formulations characterize the ordinal distribution in terms of the discrete hazard, so are natural in time to event applications (Albert and Chib, 2001).

8.2 Probit Models

Logistic models are often preferred over probit models due to the more intuitive interpretation of their regression coefficients in terms of odds ratios; however, it is worth noting that our approach for normal models is easily modified to accomplish model selection for probit mixed models by applying the well-known data augmentation Gibbs sampler described in Albert and Chib (1993). For example, using a binary response probit model of the form $P(y_{ij} = 1) = \Phi(\boldsymbol{X}'_{ij}\boldsymbol{\beta} + \boldsymbol{Z}'_{ij}\boldsymbol{a}_i)$, we introduce a latent variable v_{ij} such that $y_{ij} = 1(v_{ij} > 0)$ and $v_{ij} \sim N(\boldsymbol{X}'_{ij}\boldsymbol{\beta} + \boldsymbol{Z}'_{ij}\boldsymbol{a}_i, 1)$, yielding a conditional posterior distribution for v_{ij} of $N(\boldsymbol{X}'_{ij}\boldsymbol{\beta} + \boldsymbol{Z}'_{ij}\boldsymbol{a}_i, 1) \cdot \{1(v_{ij} > 0)y_{ij} + 1(v_{ij} < 0)(1 - y_{ij})\}$. After updating v_{ij}, the MCMC algorithm proceeds as in the normal case, except that $\sigma^2 = 1$. In our simulations this algorithm exhibited good mixing and convergence properties. This algorithm could also be adapted for ordinal probit models as described in Sect. 8.1

9 Discussion

The Bayesian framework for model selection with mixed effects models discussed here is advantageous in that it allows for fixed and random effects to be selected simultaneously. Additionally, it allows for marginal posterior inclusion probabilities to be computed for each predictor along with model-averaged coefficient estimates. Posterior model probabilities can be used to compare models; whereas frequentist testing for variance components is more limited.

In addition to model selection and averaging, the proposed prior structure and computational algorithm should be useful for efficient Gibbs sampling for fitting single mixed effects models. In particular, the prior and computational algorithm represent a useful alternative to approaches that rely on inverse-Wishart priors for variance components (e.g. Gilks et al., 1993). There is an increasing realization that inverse-Wishart priors are a poor choice, particularly when limited prior information is available. Although, we have focused on LMEs of the Laird and Ware (1982) type, it is straightforward to adapt our methods for a broader class of linear mixed models, accommodating varying coefficient models, spatially correlated data, and other applications (Zhao et al., 2006).

Gibbs sampling chains from random effects model parameters tend to exhibit slow mixing and convergence. Gelfand et al. (1996) recommend hierarchical centering for improved convergence and posterior surface behavior. Vines et al. (1994) also propose a transformation of random effects to improve mixing. A challenge in implementing the hierarchically centered model is to efficiently update the correlation matrix in the context of random effects selection where we are interested in separating out the variances. One solution is proposed by Chib and Greenberg (1998); however, it is prohibitively slow for more than a couple random effects. Further work is needed to develop fast approaches that can be easily implemented and incorporated into software packages.

Acknowledgements This work was supported in part by the Intramural Research Program of the NIH, National Institute of Environmental Health Sciences, and by the National Science Foundation under Agreement No. DMS-0112069 with the Statistical and Mathematical Sciences Institute (SAMSI). The authors are grateful to Merlise Clyde and Abel Rodriguez at Duke University and to the participants of the Model Uncertainty working group of the SAMSI Program on Latent Variable Modeling in the Social Sciences.

References

Agresti, A. (1990). *Categorical Data Analysis*. New York: Wiley

Albert, J. H. and Chib, S. (1993). Bayesian analysis of binary and polychotomous response data. *Journal of the American Statistical Association* **88**, 669–679

Albert, J. H. and Chib, S. (2001). Sequential ordinal modeling with applications to survival data. *Biometrics* **57**, 829–836

Berger, J. O., Ghosh, J. K. and Mukhopadhyay, N. (2003). Approximations and consistency of Bayes factors as model dimension grows. *Journal of Statistical Planning and Inference* **112**, 241–258

Berger, J. O. and Pericchi, L. R. (2001). Objective Bayesian methods for model selection: Introduction and comparison. In Lahiri, P., editor, *Model Selection*, volume 38 of *IMS Lecture Notes – Monograph Series*, pages 135–193. Institute of Mathematical Statistics

Breslow, N. (2003). Whither PQL? *UW Biostatistics Working Paper Series* Working Paper 192

Breslow, N. and Clayton, D. (1993). Approximate inference in generalized linear mixed models. *Journal of the American Statistical Association* **88**, 9–25

Cai, B. and Dunson, D. B. (2006). Bayesian covariance selection in generalized linear mixed models. *Biometrics* **62**, 446–457

Chen, Z. and Dunson, D. B. (2003). Random effects selection in linear mixed models. *Biometrics* **59**, 762–769

Chib, S. (1995). Marginal likelihood from the Gibbs output. *Journal of the American Statistical Association* **90**, 1313–1321

Chib, S. and Greenberg, E. (1998). Bayesian analysis of multivariate probit models. *Biometrika* **85**, 347–361

Chung, Y. and Dey, D. (2002). Model determination for the variance component model using reference priors. *Journal of Statistical Planning and Inference* **100**, 49–65

Clyde, M. and George, E. I. (2004). Model uncertainty. *Statistical Science* **19**, 81–94

Fernandez, C., Ley, E. and Steel, M. F. (2001). Benchmark priors for Bayesian model averaging. *Journal of Econometrics* **100**, 381–427

Foster, D. P. and George, E. I. (1994). The risk inflation criterion for multiple regression. *Annals of Statistics* **22**, 1947–1975

Gelfand, A. E. and Smith, A. (1990). Sampling-based approaches to calculating marginal densities. *Journal of the American Statistical Association* **85**, 398–409

Gelfand, A. E., Sahu, S. K. and Carlin, B. P. (1996). Efficient parameterizations for generalized linear mixed models. *Bayesian Statistics* **5**,

Gelman, A. (2005). Prior distributions for variance parameters in hierarchical models. *Bayesian Analysis* **1**, 1–19

George, E. I. and McCulloch, R. E. (1993). Variable selection via Gibbs sampling. *Journal of the American Statistical Association* **88**, 881–889

George, E. I. and McCulloch, R. E. (1997). Approaches for Bayesian variable selection. *Statistica Sinica* **7**, 339–374

Gerlach, R., Bird, R. and Hall, A. (2002). Bayesian variable selection in logistic regression: Predicting company earnings direction. *Australian and New Zealand Journal of Statistics* **42**, 155–168

Geweke, J. (1996). Variable selection and model comparison in regression. In *Bayesian Statistics 5 – Proceedings of the Fifth Valencia International Meeting*, pages 609–620

Gilks, W., Wang, C., Yvonnet, B. and Coursaget, P. (1993). Random-effects models for longitudinal data using Gibbs sampling. *Biometrics* **49**, 441–453

Green, P. J. (1995). Reversible jump Markov chain Monte Carlo computation and Bayesian model determination. *Biometrika* **82**, 711–732

Green, P. J. (1997). Discussion of "The EM algorithm – an old folk song sung to a fast new tune," by Meng and van Dyk. *Journal of the Royal Statistical Society, Series B* **59**, 554–555

Hall, D. and Praestgaard, J. (2001). Order-restricted score tests for homogeneity in generalised linear and nonlinear mixed models. *Biometrika* **88**, 739–751

Hastings, W. (1970). Monte Carlo sampling methods using Markov chains and their applications. *Biometrika* **57**, 97–109

Hobert, J. P. and Casella, G. (1996). The effect of improper priors on Gibbs sampling in hierarchical linear mixed models. *Journal of the American Statistical Association* **91**, 1461–1473

Holmes, C. and Knorr-Held, L. (2003). Efficient simulation of Bayesian logistic regression models. Technical report, Ludwig Maximilians University Munich

Jang, W. and Lim, J. (2005). Estimation bias in generalized linear mixed models. Technical report, Institute for Statistics and Decision Sciences, Duke University

Jiang, J., Rao, J., Gu, Z. and Nguyen, T. (2008). Fence methods for mixed model selection. *Annals of Statistics*, to appear

Johnson, V. E. and Albert, J. H. (1999). *Ordinal Data Modeling*. Berlin Heidelberg New York: Springer

Kass, R. E. and Wasserman, L. (1995). A reference Bayesian test for nested hypotheses and its relationship to the schwarz criterion. *Journal of the American Statistical Association* **90**, 928–934

Kinney, S. K. and Dunson, D. B. (2007). Fixed and random effects selection in linear and logistic models. *Biometrics* **63**, 690–698

Laird, N. and Ware, J. (1982). Random-effects models for longitudinal data. *Biometrics* **38**, 963–974

Liang, F., Paulo, R., Molina, G., Clyde, M. A. and Berger, J. O. (2005). Mixtures of g-priors for Bayesian variable selection. Technical Report 05-12, ISDS, Duke University

Lin, X. (1997). Variance component testing in generalised linear models with random effects. *Biometrika* **84**, 309–326

Liu, J. S. and Wu, Y. N. (1999). Parameter expansion for data augmentation. *Journal of the American Statistical Association* **94**, 1264–1274

Liu, C., Rubin, D. B. and Wu, Y. N. (1998). Parameter expansion to accelerate EM: the PX-EM algorithm. *Biometrika* **85**, 755–770

Meng, X. L. and Wong, W. H. (1996). Simulating ratios of normalizing constants via a simple identity: a theoretical exploration. *Statistica Sinica* **6**, 831–860

Mira, A. and Tierney, L. (2002). Efficiency and convergence properties of slice samplers. *Scandinavian Journal of Statistics* **29**, 1–12

Mitchell, T. J. and Beauchamp, J. J. (1988). Bayesian variable selection in linear regression (with discussion). *Journal of the American Statistical Association* **83**, 1023–1036

Neal, R. M. (2000). Slice sampling. Technical report, Department of Statistics, University of Toronto

Newton, M. and Raftery, A. E. (1994). Approximate Bayesian inference by the weighted likelihood bootstrap (with discussion). *Journal of the Royal Statistical Society, Series B* **56**, 3–48

O'Brien, S. M. and Dunson, D. B. (2004). Bayesian multivariate logistic regression. *Biometrics* **60**, 739–746

Pauler, D. K., Wakefield, J. C. and Kass, R. E. (1999). Bayes factors and approximations for variance component models. *Journal of the Americal Statistical Association* **94**,1242–1253

Raftery, A. E. (1995). Bayesian model selection in social resarch. *Sociological Methodology* **25**, 111–163

Schwarz, G. (1978). Estimating the dimension of a model. *Annals of Statistics* **6**, 461–464

Sinharay, S. and Stern, H. S. (2001). Bayes factors for variance component testing in generalized linear mixed models. In *Bayesian Methods with Applications to Science, Policy, and Official Statistics*, pages 507–516

Smith, M. and Kohn, R. (1996). Nonparametric regression using Bayesian variable selection. *Journal of Econometrics* **75**, 317–343

Stiratelli, R., Laird, N. M. and Ware, J. H. (1984). Random-effects model for several observations with binary response. *Biometrics* **40**, 961–971

Verbeke, G. and Molenberghs, G. (2003). The use of score tests for inference on variance components. *Biometrics* **59**, 254–262

Vines, S., Gilks, W. and Wild, P. (1994). Fitting Bayesian multiple random effects models. Technical report, Biostatistics Unit, Medical Research Council, Cambridge

West, M. (1987). On scale mixtures of normal distributions. *Biometrika* **74**, 646–648

Zellner, A. and Siow, A. (1980). Posterior odds ratios for selected regression hypotheses. In *Bayesian Statistics: Proceedings of the First International Meeting held in Valencia (Spain)*

Zhao, Y., Staudenmayer, J., Coull, B. and Wand, M. (2006). General design Bayesian generalized linear mixed models. *Statistical Science* **21**, 35–51

Bayesian Variable Selection in Generalized Linear Mixed Models

Bo Cai and David B. Dunson

1 Introduction

1.1 Background and Motivation

Repeated measures and longitudinal data are commonly collected for analysis in epidemiology, clinical trials, biology, sociology, and economic sciences. In such studies, a response is measured repeatedly over time for each subject under study, and the number and timing of the observations often varies among subjects. In contrast to cross-sectional studies that collect a single measurement per subject, longitudinal studies have the extra complication of within-subject dependence in the repeated measures. Such dependence can be thought to arise due to the impact of unmeasured predictors. Main effects of unmeasured predictors lead to variation in the average level of response among subjects, while interactions with measured predictors lead to heterogeneity in the regression coefficients. This justification has motivated random effects models, which allow the intercept and slopes in a regression model to be subject-specific. Random effects models are broadly useful for modeling of dependence not only for longitudinal data but also in multicenter studies, meta analysis and functional data analysis.

The chapter by Kinney and Dunson has motivated and described a Bayesian approach for selecting fixed and random effects in linear and logistic mixed effects models. The goal of the current chapter is to outline a Bayesian methodology for solving the same types of problems in the broader class of generalized linear mixed

B. Cai
Department of Epidemiology and Biostatistics, Arnold School of Public Health, University of South Carolina
bocai@gwm.sc.edu

D.B. Dunson
Biostatistics Branch, National Institute of Environmental Health Sciences Research Triangle Park, NC 27709, U.S.A

D. B. Dunson (ed.) *Random Effect and Latent Variable Model Selection*,
DOI: 10.1007/978-0-387-76721-5,

models (GLMMs). GLMMs provide an extension of generalized linear models (GLMs) to accommodate correlation and allow rich classes of distributions through allowing subject-specific regression coefficients in a GLM (McCulloch and Searle, 2001). Typically these subject-specific coefficients, or random effects, are assumed to have a multivariate Gaussian distribution a priori, as will be the focus in this chapter. For a recently proposed approach that allows the random effects distribution to be unknown, while also allowing fixed and random effects selection, refer to Cai and Dunson (2007).

Note that GLMMs typically assume that the observations are conditionally independent given the random effects. However, in marginalizing out the random effects, a dependence structure is induced in the multiple responses from a subject. In addition, random effects can be incorporated in GLMs to allow richer classes of distributions. For example, by incorporating random effects in Poisson or binomial GLMs, one induces over-dispersion. The resulting marginal distributions are no longer Poisson or binomial, but are instead mixtures of Poisson or binomial distributions. The form of the link function can also be impacted. For example, in marginalizing out random effects in a logistic regression model, one induces a logistic-normal link. Hence, GLMMs are often useful even when there is a single observation per subject, and modeling of dependence is not of interest. In performing inferences and variable selection in GLMMs, it is important to keep in mind the dual role of the random effects in inducing a more flexible class of models for a single outcome and in accommodating dependence in repeated outcomes. Such a duality does not occur in normal linear mixed effect models, since one still obtains a normal linear model in marginalizing out the random effects.

In addition to the complication in interpretation arising from this duality, GLMMs are certainly more complicated to fit than linear mixed models or GLMs without random effects. The challenges in fitting GLMMs arise because the marginal likelihood obtained in integrating out the random effects is not available analytically except in the normal linear model special case. Hence, one cannot obtain a simple iterative solution for maximizing the exact marginal likelihood, and even Bayesian MCMC-based approaches tend to be more difficult to implement efficiently. There is a vast literature on frequentist and Bayesian methods for addressing this problem, for example, refer to Schall (1991), Zeger and Karim (1991), Breslow and Clayton (1993), McGilchrist (1994), and McCulloch (1997).

Such methods allow one to fit a single GLMM and to perform inferences on the fixed effect regression coefficients. Much of the literature has argued that the fixed effects are of primary interest, with the random effects incorporated as nuisance parameters to account for the complication of within-subject dependence. Frequentist methods for fitting of GLMMs tend to provide only a point estimate for the random effects covariance, but if this covariance is not of interest, such an estimate is more than sufficient. However, it is hard to think of a study in which it is only of interest to assess the effects of the predictors for typical subjects, without having an interest also in how much they vary in the predictor effects. For example, in assessing the efficacy of a drug therapy in a clinical trial, is it really the case that interest focuses only on the average effectiveness of the drug and not on how much this effectiveness varies among patients? Certainly, clinicians and patients may view a drug very differently under the following two scenarios: (A) the drug has no effect or a mild adverse effect for 50% of the patients and a dramatic beneficial effect for a small subset of the patients; or (B) the drug has an identical effect for all patients. Scenarios (A) and (B) are distinguished by the magnitude of the random effect variance.

In many applications, the primary interest is in assessing whether the random effects variance is equal to zero or not. This is often the case in genetic studies in which one wishes to test whether disease risk varies across families. However, even beyond such specialized studies, we would argue that the typical scenario faced in analysis of data with a GLMM is as follows: The study collects data for a number of different covariates and it is not known with certainty a priori which predictors should be included in the fixed effects component and which should be included in the random effects component. Hence, to appropriately allow for uncertainty in specifying the model, it would be appealing to consider a Bayesian approach for modeling averaging and selection. In addition, it is typically the case that investigators desire a weight of evidence that a particular predictor is in the fixed and/or random effects component. Our goal is to describe an approach that simultaneously searches a potentially large model space for good subsets of predictors to include in the two components, while also estimating marginal inclusion probabilities and allowing model-averaged predictions.

1.2 Time to Pregnancy Application

As motivation consider the application to reproductive epidemiology studies of occupational exposures. To assess the impact of a potentially adverse exposure on fecundability, the probability of conception in a noncontracepting menstrual cycle, epidemiologists commonly measure time to pregnancy (TTP). In retrospective studies, TTP is typically defined as the number of menstrual cycles during which the woman was having intercourse without contraception prior to her most recent pregnancy.

Because TTP is a discrete event time, one can consider a discrete hazards model of the form

$$\text{logit}\,\{\Pr(T_i = t \mid T_i \geq t, \mathbf{x}_{it}, \mathbf{z}_{it})\} = \mathbf{x}_{it}'\boldsymbol{\beta} + \mathbf{z}_{it}'\boldsymbol{\zeta}_i, \tag{1}$$

where T_i is the TTP for woman i, $\mathbf{x}_{it}$ and $\mathbf{z}_{it}$ are vectors of predictors that may vary from cycle to cycle, $\boldsymbol{\beta}$ are fixed effects regression coefficients, $\boldsymbol{\zeta}_i \sim \mathrm{N}(\mathbf{0}, \boldsymbol{\Sigma})$ are random effects for woman i, and $\boldsymbol{\Sigma}$ is a covariance matrix. Model (1) uses a logistic mixed effects model to characterize the conditional probability of a pregnancy in the tth cycle at risk given that the woman did not conceive prior to that cycle. If there was no unexplained heterogeneity in fecundability after accounting for the measured predictors, then the random effects could be excluded. In this case, in the absence of time-varying predictors, TTP is geometrically distributed. Over-dispersion relative to the geometric distribution allows one to identify the random effects variance even with a single TTP measurement from a woman.

Rowland et al. (1992) studied factors related to fecundability in dental assistants. Study participants completed a demographic and exposure history questionnaire, while also providing information on the number of menstrual cycles during which the woman was having noncontracepting sexual intercourse before the most recent pregnancy. Model (1) could be fitted easily in standard software packages (e.g., SAS or WinBUGS) if the predictors to be included in the fixed and random effects components were known. However, it is of course not known a priori which predictors should be included, and we would like to investigate which factors vary in their effects across women. For example, do the effects of aging, recent oral contraceptive use, and smoking vary?

1.3 Background on Model Selection in GLMMs

If the focus were on selecting the single best GLMM from among the possible candidates, one could potentially fit the model for all possible choices of fixed effect predictors, $\mathbf{x}_{it}$, and random effects predictors, $\mathbf{z}_{it}$. One could then apply a standard criterion, such as the Akaike's information criterion (AIC) or the Bayesian information criterion (BIC). However, it is not clear that these criteria are appropriate in mixed effects models, as one may need to estimate an effective degrees of freedom to provide a more appropriate penalty for model complexity. In hierarchical models, such as GLMMs, the number of parameters is arbitrary, as one can write different formulations of the same model that have different numbers of parameters.

An additional issue is that one may require a weight of evidence that a particular predictor is included in the random effects component. This can be addressed by conducting hypothesis tests of whether the variance of the random effects distribution is equal to zero. The chapter by Ciprian Crainiceanu considers likelihood ratio tests in this setting, while the chapter by Daowen Zhang and Xihong Lin considers score tests. These approaches can be used to obtain p-values for testing whether variance components are equal to zero.

In the Bayesian literature, Albert and Chib (1997) proposed an approach for testing whether a random intercept should be included, Sinharay and Stern (2001) developed a more general approach for calculating Bayes factors for variance components testing in GLMMs, and Chen et al. (2003) proposed a class of informative priors for model selection in GLMMs. These methods focus on comparing two models at a time, and do not provide a general approach for searching for promising subsets of candidate predictors.

In the setting of linear mixed models for normal data, Chen et al. (2003) proposed a Bayesian approach for random effects selection based on using variable selection priors for the components in a special decomposition of the random effects covariance. Related approaches have been used in graphical (or covariance structure) modeling for multivariate normal data (refer to Wong et al. (2003); Liechty et al. (2004) for recent references). Bayesian variable selection in conventional GLMs has also received a lot of interest in the literature. Raftery (1996) proposed an approximate

Bayes factor approach, Meyer and Laud (2002) considered predictive variable selection, Nott and Leonte (2004) developed an innovative sampling algorithm and Ntzoufras et al. (2003) developed methods for joint variable and link selection.

Cai and Dunson (2006) described a Bayesian approach to the problem of simultaneous selection of fixed and random effects in GLMMs. In this chapter, we summarize this work and provide more details of the approach. We first choose variable selection-type mixture priors for the fixed effects regression coefficients and the parameters in a special Cholesky decomposition of the random effects covariance proposed by Chen and Dunson (2003). These priors allow fixed effects to drop out of the model by placing probability mass on $\beta_l = 0$. In addition, following a related approach to Albert and Chib (1997) and Chen and Dunson (2003) (refer also to the Kinney and Dunson chapter), we assign positive probability to random effects having 0 variance to effectively move between the full model with random effects for every predictor and submodels excluding one or more random effects. This prior specification has convenient computational properties, which is important, given the potentially large number of models under consideration.

Outside of simple models, it is typically the case that Bayesian model selection requires the calculation of normalizing constants, which do not have closed forms. Unfortunately, typical MCMC algorithms bypass calculation of normalizing constants, so are not appropriate. In addition, MCMC-based methods for calculating normalizing constants tend to be highly computationally-intensive, even when considering a single model instead of a high-dimensional list of models. For these reasons, many approaches rely on analytic approximations to intractable integrals using Laplace and other approaches based on Taylor series. The Cai and Dunson (2006) approach reviewed in this chapter relies on stochastic search variable selection (SSVS) (George and McCulloch (1993)) implemented with MCMC, combined with limited analytic approximations. Similar ideas were implemented previously by Raftery et al. (1996) in the context of model averaging in survival analysis, and Chipman et al. (2002, 2003) in implementing analyses of treed GLMs.

The remainder of this chapter is organized as follows. Section 2 reviews the specification of a GLMM, and describes a Bayesian formulation of the model selection problem. Section 3 outlines the SSVS algorithm for posterior computation and model search. Section 4 considers simulation examples as a proof of concept and illustration. Section 5 applies the approach to the Rowland et al. (1992) time to pregnancy application, and Sect. 6 contains a discussion.

2 Bayesian Subset Selection in GLMMs

2.1 *Generalized Linear Mixed Models*

For observation j ($j = 1, \ldots, n_i$) from subject i ($i = 1, \ldots, n$), let y_{ij} denote the response variable, let $\mathbf{x}_{ij}$ denote a $p \times 1$ vector of candidate predictors, and let $\mathbf{z}_{ij}$ denote a $q \times 1$ vector of candidate predictors. Note that it is important to distinguish candidate predictors from predictors that are included in the model. Here, we will follow the approach of imbedding all of the models under consideration in a full model that contains all of the candidate predictors. Then, by choosing a prior that allows the fixed effect coefficients to have values exactly equal to zero, we allow each of the fixed effect candidate predictors to potentially drop out of the model. In addition, by choosing a prior that allows the random effect variances to be exactly zero, we allow predictors to drop out of the random effect component.

Reviewing the GLMM specification, note that the elements of $\mathbf{y}_i = (y_{i1}, \ldots, y_{i,n_i})'$ are modeled as conditionally-independent random variables from a simple exponential family

$$\pi(y_{ij} \mid \mathbf{x}_{ij}, \mathbf{z}_{ij}, \boldsymbol{\zeta}_i) = \exp\left\{\frac{y_{ij}\theta_{ij} - b(\theta_{ij})}{a_{ij}(\phi)} + c(y_{ij}, \phi)\right\}, \tag{2}$$

where θ_{ij} is canonical parameter related to the linear predictor $\eta_{ij} = \mathbf{x}_{ij}'\boldsymbol{\beta} + \mathbf{z}_{ij}'\boldsymbol{\zeta}_i$ with a $p \times 1$ vector of fixed effects regression coefficients $\boldsymbol{\beta}$, and a $q \times 1$ vector of subject-specific random effects $\boldsymbol{\zeta}_i \sim \mathrm{N}_q(\mathbf{0}, \boldsymbol{\Sigma})$, ϕ is a scalar dispersion parameter, and $a_{ij}(\cdot), b(\cdot), c(\cdot)$ are known functions, with $a_{ij}(\phi)$ typically expressed as ϕ/w_{ij}, where w_{ij} is a known weight.

Note that the conditional-independence assumption implies that the dependence among the different observations from subject i arises entirely from the shared dependence on the random effects. In marginalizing out the random effects, one induces a predictor-dependent correlation structure in the multivariate response vector, $\mathbf{y}_i$. For this reason, GLMMs provide a very useful class of models for modeling of multivariate non-normal data. In addition, in selecting predictors to include in the random effects component, we are also simultaneously selecting a covariance structure among the multiple outcomes.

Let $\mathbf{y} = (\mathbf{y}_1', \ldots, \mathbf{y}_n')'$, $\mathbf{y}_i = (y_{i1}, \ldots, y_{in_i})'$, $\mathbf{X} = (\mathbf{X}_1', \ldots, \mathbf{X}_n')'$, $\mathbf{X}_i = (\mathbf{x}_{i1}, \ldots, \mathbf{x}_{in_i})'$, $\mathbf{Z} = \mathrm{diag}(\mathbf{Z}_1, \ldots, \mathbf{Z}_n)$, $\mathbf{Z}_i = (\mathbf{z}_{i1}, \ldots, \mathbf{z}_{in_i})'$, and $\boldsymbol{\zeta} = (\boldsymbol{\zeta}_1', \ldots, \boldsymbol{\zeta}_n')'$. The joint distribution of responses $\mathbf{y}$ and random effects $\boldsymbol{\zeta}$ conditionally on the predictors $\mathbf{X}$ and $\mathbf{Z}$ is of the form

$$\pi(\mathbf{y}, \boldsymbol{\zeta} \mid \boldsymbol{\beta}, \phi, \boldsymbol{\Sigma}, \mathbf{X}, \mathbf{Z}) = \exp\left[\left\{\mathbf{y}'h(\boldsymbol{\eta}) - \mathbf{b}'(h(\boldsymbol{\eta}))\,\mathbf{1}_N\right\}/\mathbf{a}' + \mathbf{c}'(\mathbf{y}, \phi)\mathbf{1}_N\right]\pi(\boldsymbol{\zeta}|\boldsymbol{\Sigma}), \tag{3}$$

where $\boldsymbol{\eta} = \mathbf{X}\boldsymbol{\beta} + \mathbf{Z}\boldsymbol{\zeta}$ and $\mathbf{1}_N$ is an $N \times 1$ vector of ones, where $N = \sum_{i=1}^n n_i$.

In practice, one needs to choose a particular exponential family distribution and link function to complete an explicit specification of the model. One aspect of model uncertainty in GLMMs is choice of the distribution and link function. However, in this chapter, we assume that both these components are known to simplify exposition.

Our focus is on Bayesian approaches for accounting for uncertainty in the elements of $\mathbf{x}_{ij}$ and $\mathbf{z}_{ij}$ to be included in the model, as well as the covariance structure in the $\boldsymbol{\zeta}_i$'s. As discussed in detail in the chapter by Kinney and Dunson, a Bayesian specification of the model uncertainty problem requires one to choose a prior on the model space, corresponding to prior probabilities for each of the models in the list, along with priors for the coefficients within each of the models. We let $\mathcal{M}$ denote the list of models corresponding to all possible subsets of $\mathbf{x}_{ij}$ and $\mathbf{z}_{ij}$. In addition, we let $M \in \mathcal{M}$ be an index for a single model in $\mathcal{M}$.

With this notation in place, we let $\mathbf{x}_{ij}^{(M)}$, $\mathbf{z}_{ij}^{(M)}$, $\boldsymbol{\beta}^{(M)}$, $\boldsymbol{\zeta}_i^{(M)}$, and $\boldsymbol{\Sigma}^{(M)}$ denote the terms for model M, which is specified as $\eta_{ij}^{(M)} = \mathbf{x}_{ij}^{(M)'}\boldsymbol{\beta}^{(M)} + \mathbf{z}_{ij}^{(M)'}\boldsymbol{\zeta}_i^{(M)}$, with the dispersion parameter, link function, and distributional form assumed common to the different models $M \in \mathcal{M}$. The predictors $\mathbf{x}_{ij}^{(M)}$ consist of a $p_M \leq p$ subset of $\mathbf{x}_{ij}$, while $\mathbf{z}_{ij}^{(M)}$ is a $q_M \leq q$ subset of $\mathbf{z}_{ij}$. In addition, $\boldsymbol{\zeta}_i^{(M)} \sim \mathrm{N}(\mathbf{0}, \boldsymbol{\Sigma}^{(M)})$ is a $q_M \times 1$ vector of random effects, with $q_M \times q_M$ covariance matrix $\boldsymbol{\Sigma}^{(M)}$, which can have zero off-diagonal elements corresponding to conditional independence relationships in the random effects included.

The model space, $\mathcal{M}$, includes all possible combinations of subsets of $\mathbf{x}_{ij}$ and $\mathbf{z}_{ij}$ and zero off-diagonal elements of the random effects covariance matrices corresponding to these subsets. Hence, the total number of models is $2^p \sum_{k=0}^{q} \binom{q}{k} 2^{\frac{1}{2}(q-k)(q-k-1)}$. Clearly, the number of models under consideration grows extremely fast with p and q, so that it would be infeasible to run a separate MCMC analysis for each model in the list even for a modest number of candidate predictors. For example, even with $p = q = 5$, we have 46,400 models in $\mathcal{M}$.

2.2 Description of Approach

Our goal is to select good models from among the different possibilities for M. To attempt to identify good models quickly from among the potentially enormous number of possible models under consideration, we apply a stochastic search variable selection approach. This algorithm sequentially modifies the variables included in each component through MCMC sampling. For the fixed effects component, we follow the common convention of choosing mixture priors with point mass at zero. For the random effects, a more innovative and involved approach is necessary. In particular, we propose to induce zero variance components and zero correlations between random effects through zeroing coefficients in a carefully chosen decomposition of the random effects covariance.

In Bayesian analyses, the standard prior for a covariance matrix is the Wishart prior. However, it is widely known that the Wishart prior is very inflexible, allowing only a single degrees of freedom for all elements and not allowing zero off-diagonal elements. Because the constraints on a covariance matrix limit the flexibility with which one can consider direct modifications to the Wishart prior, a common trick is to induce a prior for a covariance matrix through priors for parameters in a decomposition. For example, Daniels and Zhao (2003) used a special Cholesky decomposition to model changes in the random effects covariance over time. A related decomposition approach was considered by Daniels and Pourahmadi (2002). Daniels and Kass (1999) instead considered spectral decomposition. Wong et al.

(2003) proposed a Bayesian method for estimating an inverse covariance matrix for normal data using a prior that allows the off-diagonal elements of the inverse covariance matrix to be zero.

Chen and Dunson (2003) proposed an alternative decomposition, which has some appealing characteristics. For example, the decomposition results in a conditionally linear regression model, which allows one to choose a conditionally-conjugate prior. This conjugacy is important in allowing closed form calculation of conditional probabilities of including a predictor. Such probabilities are required in implementing SSVS algorithms. Unlike Chen and Dunson (2003) and Kinney and Dunson (2007), this chapter will allow zero off-diagonal elements in the random effects covariance matrix. This is accomplished through variable selection mixture priors for parameters in the decomposition that control correlations among the random effects. Effectively, the prior allows movement between models of different dimension, with the covariance matrix of the random effects in each of these models being positive semidefinite. The details are given Sect. 2.3.

2.3 Reparameterization and Mixture Prior Specification

The random effects covariance matrix $\mathbf{\Sigma}$ may be factorized as

$$\mathbf{\Sigma} = \mathbf{\Lambda}\mathbf{\Gamma}\mathbf{\Gamma}'\mathbf{\Lambda},$$

where $\mathbf{\Lambda} = \text{diag}(\lambda_1, \ldots, \lambda_q)$ is a diagonal matrix, with diagonal elements $\lambda_k \geq 0$ for $k = 1, \ldots, q$, and $\mathbf{\Gamma}$ denotes the lower triangular matrix

$$\begin{pmatrix} 1 & & & \\ \gamma_{21} & 1 & & \\ \vdots & \vdots & \ddots & \\ \gamma_{q,1} & \gamma_{q,2} & \cdots & 1 \end{pmatrix}.$$

From straightforward algebra, this decomposition of the random effects covariance implies that the (k, l) element of the matrix Σ has the following expression:

$$\sigma_{kl} = \sigma_{lk} = \lambda_k\lambda_l\left(\gamma_{r_2,r_1} + \sum_{s=1}^{r_1-1} \gamma_{ks}\gamma_{ls}\right), \quad \text{for } k, l = 1, \ldots, q, \tag{4}$$

where $r_1 = \min(k, l)$, $r_2 = \max(k, l)$. Hence, the λs are row and column-specific multipliers, while the γ's control the size of the off-diagonal elements of Σ. For example, in the special case in which all the lower triangular elements of Γ's equal zero, Σ is a diagonal matrix with λ_k^2, for $k = 1, \ldots, q$ the elements along the diagonal. One obtains a positive semidefinite $\mathbf{\Sigma}$ when $\lambda_k > 0$ for all k.

Because λ_k serves as a multiplier on all the elements in the kth row and column of Σ, it is clear that one effectively excludes the kth random effect from the model when $\lambda_k = 0$, as in this case the random effects variance will equal zero. Note that once all the rows and columns corresponding to null random effects having zero variance are excluded, one obtains a positive semidefinite covariance matrix for those random effects included in the model. In this way, random effects are allowed to effectively drop out of the model.

Recall that, to complete a Bayesian formulation of the model selection problem, we need to choose prior probabilities for each $M \in \mathcal{M}$, along with prior distributions for the coefficients within each model. Assuming that a common prior is assumed for the parameters that belong to every model in the list, such as the exponential family dispersion parameter, we focus on the choice of prior for $\boldsymbol{\lambda} = (\lambda_1, \ldots, \lambda_q)'$ and $\boldsymbol{\gamma} = (\gamma_{mk} : m = k+1, \ldots, q; k = 1, \ldots, q-1)'$. Each model $M \in \mathcal{M}$ is distinguished by the subsets of $\boldsymbol{\lambda}$ and $\boldsymbol{\gamma}$ having zero elements. Hence, by choosing a single mixture prior for $\boldsymbol{\lambda}$ and $\boldsymbol{\gamma}$ that allows for zero elements, we simultaneously specify a prior over the model space $\mathcal{M}$ and for the coefficients within each model.

To drop out the off-diagonal elements in the covariance matrix when a random effect is excluded, the support of the prior for $\boldsymbol{\gamma}$ is defined as $\mathcal{R}_{\boldsymbol{\lambda}} = \{\boldsymbol{\gamma} : \gamma_{mk} = \gamma_{kl} = 0 \text{ if } \lambda_k = 0, \text{ for } k = 1, \ldots, q, 1 \leq l < k < m \leq q, l, m \in \mathcal{N}\}$. Since the covariance matrix $\boldsymbol{\Sigma}$ is a function of $\boldsymbol{\lambda}$ and $\boldsymbol{\gamma}$, the prior density of $\boldsymbol{\Sigma}$ is induced through the priors for $\boldsymbol{\lambda}$ and $\boldsymbol{\gamma}$, $\pi(\boldsymbol{\lambda}, \boldsymbol{\gamma}) = \pi(\boldsymbol{\gamma}|\boldsymbol{\lambda})\pi(\boldsymbol{\lambda})$. The prior for $\boldsymbol{\lambda}$ is $\prod_{k=1}^{q} \pi(\lambda_k)$, where $\pi(\lambda_k)$ is chosen as mixtures of point masses at zero and a truncated normal density

$$\pi(\lambda_k) = \pi_{1,k0} 1(\lambda_k = 0) + (1 - \pi_{1,k0}) 1(\lambda_k > 0) \frac{\mathrm{N}(\lambda_k; \mu_{1,k0}, \sigma_{1,k0}^2)}{\mathrm{F}(0; -\mu_{1,k0}, \sigma_{1,k0}^2)}, \tag{5}$$

where $\pi_{1,k0}$, $\mu_{1,k0}$ and $\sigma_{1,k0}^2$ are hyperparameters specified by investigators, and $\mathrm{F}(\cdot)$ is the normal distribution function. We refer to prior (5) as a zero-inflated positive normal density, $\text{ZI-N}^+(\lambda_k; \pi_{1,k0}, \mu_{1,k0}, \sigma_{1,k0}^2)$. The prior probability of the kth random effect being excluded is $\pi_{1,k0} = \Pr(H_{0k} : \lambda_k = 0)$. The prior probability of the global null hypothesis of homogeneity is $\Pr(H_0 : \lambda_1 = \cdots = \lambda_q = 0) = \prod_{k=1}^{q} \pi_{1,k0}$, which implies that all random effects are excluded from the model.

To allow fixed effects predictors to effectively drop out of the model, we also choose a zero-inflated normal density, $\text{ZI-N}(\beta_v | \pi_{2,v0}, \mu_{2,v0}, \sigma_{2,v0}^2)$, as the prior for β_v, for $v = 1, \ldots, p$. The prior probability of the vth predictor being excluded is then $\pi_{2,v0} = \Pr(\beta_v = 0)$. Similar mixture priors have been widely used in the Bayesian variable selection literature (cf. Geweke, 1996).

We also allow zero off-diagonal elements in the covariance matrix by choosing mixture priors with masses at 0 for the $\boldsymbol{\gamma}$'s. We choose a zero-inflated normal density, $\text{ZI-N}(\gamma_{mk}; \pi_{3,mk,0}, \mu_{3,mk,0}, \sigma_{3,mk,0}^2)$, with the constraint related to $\boldsymbol{\lambda}$, as the prior for γ_{mk}, for $m = k+1, \ldots, q$ and $k = 1, \ldots, q-1$. This mixture prior fixes the prior probability of $\gamma_{mk} = 0$ to be $\pi_{3,mk,0}$. In this way, the correlations between the random effects can be zero or nonzero. Explicitly, from (4), the correlation coefficient between the mth and the kth random effects is

$$\rho(\zeta_{im}, \zeta_{ik}; \boldsymbol{\gamma}) = \frac{\gamma_{mk} + \sum_{s=1}^{k-1} \gamma_{ks}\gamma_{ms}}{\sqrt{\left(1 + \sum_{s=1}^{m-1} \gamma_{ms}^2\right)\left(1 + \sum_{s=1}^{k-1} \gamma_{ks}^2\right)}}.$$

So the prior probability that the two random effects are uncorrelated is

$$\Pr\{\rho(\zeta_{im}, \zeta_{ik}) = 0\} = \Pr(\gamma_{m1}\gamma_{k1} = \cdots = \gamma_{m,k-1}\gamma_{k,k-1} = \gamma_{mk} = 0)$$
$$= \pi_{3,mk,0} \prod_{s=1}^{k-1} \left\{\pi_{3,ms,0}(1 - \pi_{3,ks,0}) + \pi_{3,ks,0}\right\}.$$

Note that the expression for the correlation coefficients, $\rho(\zeta_{im}, \zeta_{ik})$, for any two random effects that have nonzero variance ($\lambda_m > 0$, $\lambda_k > 0$) does not involve $\boldsymbol{\lambda}$.

Even though we use the λs and γ s in specifying the prior and in posterior computation, inferences should be based on the random effect variances and correlations, which are easily calculated from the λs and γ s. To obtain samples from the prior distribution of $\boldsymbol{\Sigma}$, one can simply draw from the prior of $\boldsymbol{\lambda}$, $\boldsymbol{\gamma}$ and then calculate the corresponding values of $\boldsymbol{\Sigma}$. Similarly to obtain samples from the posterior distribution of $\boldsymbol{\Sigma}$, one can rely on samples from the posterior of $\boldsymbol{\lambda}$, $\boldsymbol{\gamma}$.

To choose the values for the hyperparameters, we suggest an informative specification. For example, one could set the point mass probabilities equal to 0.5 to allow equal probability of inclusion or exclusion, while centering the priors for the coefficients on zero, and choosing the prior variance to assign high probability to a wide range of plausible values for the random effects covariance values. However, it is important to avoid choose very high variance, diffuse but proper priors. Diffuse priors are not recommended for Bayesian model selection, because the higher the prior variance the more the null model is favored. Default prior selection in GLMMs in an interesting area for future research.

2.4 An Approximation

Our goal is to implement a stochastic search variable selection (SSVS) algorithm for simultaneously exploring the model space, $\mathcal{M}$, while also obtaining draws from the posterior distributions for the parameters. If we had a linear mixed effects model, then the steps in the SSVS algorithm, obtained from the specification in Sects. 2.1–2.3, would all involve sampling from standard distributions. However, in GLMMs more broadly, this is no longer the case, and we cannot apply standard MCMC algorithms for updating the parameters in a single GLMM directly. The problem occurs in calculating the conditional posterior distributions for a single element of $\boldsymbol{\lambda}$ or $\boldsymbol{\gamma}$. In particular, due to the use of the variable selection mixture prior, the conditional posterior is also a mixture of a point mass at zero and a continuous distribution. Calculation of the conditional posterior probability allocated to the continuous component involves calculating a marginal likelihood, which is not available in closed form.

In particular, it is necessary to calculate the marginal likelihood of $\mathbf{y}$ conditional on the parameters by integrating out the random effects

$$L(\boldsymbol{\beta}, \phi, \boldsymbol{\Sigma}; \mathbf{y}, \mathbf{X}, \mathbf{Z}) = \int_{\Re^q} \pi(\mathbf{y}|\boldsymbol{\beta}, \phi, \boldsymbol{\zeta}, \mathbf{X}, \mathbf{Z})\pi(\boldsymbol{\zeta}|\boldsymbol{\Sigma})d\boldsymbol{\zeta}. \tag{6}$$

Let $l(\boldsymbol{\beta}, \phi, \boldsymbol{\Sigma}; \mathbf{y}) = \log L(\boldsymbol{\beta}, \phi, \boldsymbol{\Sigma}; \mathbf{y})$, suppressing the conditioning on $\mathbf{X}$ and $\mathbf{Z}$ as shorthand. By far the most commonly-used and successful approach for analytically approximating marginal likelihoods is the Laplace approximation (Solomon and Cox, 1992; Breslow and Clayton (1993); Lin, 1997; Chipman et al., 2003, among others). Recall that $\boldsymbol{\Sigma}$ depends on $\boldsymbol{\lambda}, \boldsymbol{\gamma}$ through expression (4). In addition, when $\boldsymbol{\lambda} = \mathbf{0}$, we have $\boldsymbol{\Sigma} = 0$, which implies that $\boldsymbol{\zeta} \equiv \mathbf{0}$ so that $L(\boldsymbol{\beta}, \phi, \boldsymbol{\Sigma}; \mathbf{y}, \mathbf{X}, \mathbf{Z}) = \pi(\mathbf{y}|\boldsymbol{\beta}, \phi, \mathbf{X})$, which is the likelihood for a GLM with no random effects.

In the general case, Cai and Dunson (2006) proposed a second-order Taylor series approximation to (6). In particular, start by approximating the first integrand of (6) by taking a second-order Taylor series expansion at $\mathrm{E}(\boldsymbol{\zeta}) = \mathbf{0}$, the mean of the random effects

$$
\begin{aligned}
L(\boldsymbol{\beta}, \phi, \boldsymbol{\zeta}; \mathbf{y}) &\approx L(\boldsymbol{\beta}, \phi, \boldsymbol{\zeta} = \mathbf{0}; \mathbf{y}) + \left.\frac{\partial L(\boldsymbol{\beta}, \phi, \boldsymbol{\zeta}; \mathbf{y})}{\partial \boldsymbol{\zeta}}\right|_{\boldsymbol{\zeta}=\mathbf{0}} \boldsymbol{\zeta} + \frac{1}{2}\boldsymbol{\zeta}' \left.\frac{\partial^2 L(\boldsymbol{\beta}, \phi, \boldsymbol{\zeta}; \mathbf{y})}{\partial \boldsymbol{\zeta} \partial \boldsymbol{\zeta}'}\right|_{\boldsymbol{\zeta}=\mathbf{0}} \boldsymbol{\zeta} \\
&= L(\boldsymbol{\beta}, \phi, \boldsymbol{\zeta} = \mathbf{0}; \mathbf{y}) \left\{ 1 + \left.\frac{\partial l(\boldsymbol{\beta}, \phi, \boldsymbol{\zeta}; \mathbf{y})}{\partial \boldsymbol{\eta}}\right|_{\boldsymbol{\zeta}=\mathbf{0}} \mathbf{Z}\boldsymbol{\zeta} \right. \\
&\quad + \frac{1}{2}(\mathbf{Z}\boldsymbol{\zeta})' \left(\left[\frac{\partial l(\boldsymbol{\beta}, \phi, \boldsymbol{\zeta}; \mathbf{y})}{\partial \boldsymbol{\eta}} \frac{\partial l(\boldsymbol{\beta}, \phi, \boldsymbol{\zeta}; \mathbf{y})}{\partial \boldsymbol{\eta}'} \right. \right. \\
&\quad \left. \left. \left. + \mathrm{DG}\left\{ \frac{\partial^2 l(\boldsymbol{\beta}, \phi, \boldsymbol{\zeta}; \mathbf{y})}{\partial \boldsymbol{\eta} \partial \boldsymbol{\eta}'} \right\} \right] \right|_{\boldsymbol{\zeta}=\mathbf{0}} \right) \mathbf{Z}\boldsymbol{\zeta} \right\},
\end{aligned}
$$

where $\boldsymbol{\eta} = \mathbf{X}\boldsymbol{\beta} + \mathbf{Z}\boldsymbol{\zeta}$, and $\mathrm{DG}(A)$ denotes a diagonal matrix with diagonal entries of A. We note that (6) is actually the expectation of $L(\boldsymbol{\beta}, \phi, \boldsymbol{\zeta}; \mathbf{y})$ with respect to $\boldsymbol{\zeta}$. Thus, the approximation $\widetilde{L}(\boldsymbol{\beta}, \phi, \boldsymbol{\Sigma}; \mathbf{y})$ can be expressed as

$$
\widetilde{L}(\boldsymbol{\beta}, \phi, \boldsymbol{\Sigma}; \mathbf{y}) = L_0 \left\{ 1 + \frac{1}{2} \mathrm{tr}\left(\mathbf{Z}' \left[\frac{\partial l(\boldsymbol{\beta}, \phi, \boldsymbol{\zeta}; \mathbf{y})}{\partial \boldsymbol{\eta}} \frac{\partial l(\boldsymbol{\beta}, \phi, \boldsymbol{\zeta}; \mathbf{y})}{\partial \boldsymbol{\eta}'} + \mathrm{DG}\left\{ \frac{\partial^2 l(\boldsymbol{\beta}, \phi, \boldsymbol{\zeta}; \mathbf{y})}{\partial \boldsymbol{\eta} \partial \boldsymbol{\eta}'} \right\} \right] \right|_{\boldsymbol{\zeta}=\mathbf{0}} \mathbf{Z}\boldsymbol{\Sigma}^* \right) \right\}, \tag{7}
$$

where $L_0 = L(\boldsymbol{\beta}, \phi, \boldsymbol{\zeta} = \mathbf{0}; \mathbf{y})$, which denotes the likelihood for the ordinary GLM, $\mathrm{tr}(A)$ denotes the trace of matrix A, and $\boldsymbol{\Sigma}^* = I_n \otimes \boldsymbol{\Sigma}$, the Kronecker product of I_n and $\boldsymbol{\Sigma}$. The second term in (7) involves the first and second derivative calculations. Thus, the approximation (7) is tractable, since the first and second derivatives of $l(\boldsymbol{\beta}, \phi, \boldsymbol{\zeta}|\mathbf{y})$ are easily obtained as follows:

$$
\begin{aligned}
\frac{\partial l(\boldsymbol{\beta}, \phi, \boldsymbol{\zeta}|\mathbf{y})}{\partial \boldsymbol{\eta}} &= \left\{ \mathbf{y} - \frac{\partial \psi(h(\boldsymbol{\eta}))}{\partial h(\boldsymbol{\eta})} \right\} \frac{\partial h(\boldsymbol{\eta})}{\phi \partial \boldsymbol{\eta}}, \\
\frac{\partial^2 l(\boldsymbol{\beta}, \phi, \boldsymbol{\zeta}|\mathbf{y})}{\partial \boldsymbol{\eta} \partial \boldsymbol{\eta}'} &= \left\{ \mathbf{y} - \frac{\partial \psi(h(\boldsymbol{\eta}))}{\partial h(\boldsymbol{\eta})} \right\} \frac{\partial^2 h(\boldsymbol{\eta})}{\phi \partial \boldsymbol{\eta} \partial \boldsymbol{\eta}'} - \frac{\partial^2 \psi(h(\boldsymbol{\eta}))}{\phi \partial h(\boldsymbol{\eta}) \partial h(\boldsymbol{\eta}')} \frac{\partial h(\boldsymbol{\eta})}{\partial \boldsymbol{\eta}} \frac{\partial h(\boldsymbol{\eta})}{\partial \boldsymbol{\eta}'}.
\end{aligned}
$$

Then, in general, the approximation $\widetilde{L}(\boldsymbol{\beta}, \phi, \boldsymbol{\Sigma}; \mathbf{y})$ may be expressed as

$$
L_0 \left\{ 1 + \frac{1}{2\phi} \left(\sum_{k=1}^{q} \sigma_{kk} \sum_{i=1}^{n} B_{i,k}^{(1)} + 2 \sum_{k=1}^{q-1} \sum_{m=k+1}^{q} \sigma_{mk} \sum_{i=1}^{n} B_{i,m,k}^{(2)} \right) \right\}, \tag{8}
$$

where $B_{i,k}^{(1)}$ and $B_{i,m,k}^{(2)}$ are functions of $\boldsymbol{\beta}$ related to response variable $\mathbf{y}$, fixed effects predictors $\mathbf{X}$, and the random effects predictors $\mathbf{Z}$, and vary for particular GLMMs. In detail, the approximation (8) may be shown as

$$L_0\left[1+\frac{1}{2\phi}\left\{\sum_{k=1}^{q}\lambda_k^2\left(1+\sum_{s=1}^{k-1}\gamma_{ks}^2\right)\sum_{i=1}^{n}B_{i,k}^{(1)}+2\sum_{k=1}^{q-1}\sum_{m=k+1}^{q}\lambda_k\lambda_m\left(\gamma_{mk}+\sum_{s=1}^{k-1}\gamma_{ks}\gamma_{ms}\right)\sum_{i=1}^{n}B_{i,m,k}^{(2)}\right\}\right]. \tag{9}$$

This form gives a general analytically tractable form for GLMMs which simplifies the subsequent computation. The general result can be applied in a straightforward manner to any particular special case (e.g., logistic regression, Poisson, log linear models, etc). The detailed marginal distributions for normal linear, logistic regression and Poisson models are provided in Appendix.

We can obtain a simpler approximation to (9) under the assumption that the elements of the random effects covariance are small enough so that the assumption $\exp(\sigma) \approx 1 + \sigma$ is warranted. In this case, the approximation becomes

$$L_0 \exp\left\{\frac{1}{2\phi}\left(\sum_{k=1}^{q}\sigma_{kk}\sum_{i=1}^{n}B_{i,k}^{(1)}+2\sum_{k=1}^{q-1}\sum_{m=k+1}^{q}\sigma_{mk}\sum_{i=1}^{n}B_{i,m,k}^{(2)}\right)\right\}. \tag{10}$$

This expression is simpler to calculate rapidly, so may have advantages in certain cases.

3 Posterior Computation

3.1 General Strategies

Relying on the approximations proposed in Sect. 2.4 only when needed to approximate marginal likelihoods integrating out random effects, this section describes the steps involved in the Cai and Dunson (2006) SSVS algorithm. For binomial and Poisson likelihoods, the scale or dispersion parameter is $\phi = 1$. For normal linear models, ϕ is σ^2, and we follow common practice in choosing a gamma prior, $\mathcal{G}(c_0, d_0)$, for σ^{-2}. The SSVS algorithm iteratively samples from the full conditional distributions of each of the parameters. For $\boldsymbol{\beta}$, $\boldsymbol{\lambda}$ and $\boldsymbol{\gamma}$, these posteriors will have a mixture structure consisting of point mass at 0 and nonconjugate distributions. In calculating the point mass probabilities, we rely on the approximation described in Sect. 2.4. To sample from the nonconjugate distribution, we use adaptive rejection Metropolis sampling (Gilks et al., 1995, 1997).

In general, if a parameter θ has a mixture prior of form $\pi(\theta) = \pi_0 1(\theta = 0) + (1 - \pi_0)1(\theta \neq 0)p(\theta)$, and the likelihood is nonconjugate (e.g., the GLMM with logit link and log link), directly sampling for θ from its full conditional distribution is rather difficult due to the intractable marginal integral for θ in calculating $\hat{\pi}$. To sample θ more efficiently, one could introduce two latent variables δ and $\tilde{\theta}$ which are linked to θ as $\theta = (1-\delta)\tilde{\theta}$, where $\delta \sim \text{Bernoulli}(\pi_0)$ and $\tilde{\theta} \sim p(\tilde{\theta})$. Thus, one can sample θ through the following steps:

- Update δ from its full conditional distribution Bernoulli$(\tilde{\pi})$, where

$$\tilde{\pi} = \frac{\pi_0}{\pi_0 + (1-\pi_0)L(\theta = \tilde{\theta}, \Theta)/L(\theta = 0, \Theta)}$$

  with $L(\cdot)$ denoting the likelihood and Θ the other parameters.
- Update $\tilde{\theta}$ for θ from its full conditional distribution $L(\tilde{\theta}, \Theta)p(\tilde{\theta})$ if $\delta = 0$, and let $\theta = 0$ otherwise.

Let $\delta_{1,k}$ denote an indicator variable which is one if the kth random effect is excluded (H_{0k}) and zero if the random effect is included (H_{1k}). Then, it is clear that the prior specification in (5) can be induced through letting $\lambda_k = (1-\delta_{1,k})\tilde{\lambda}_k$, where $\delta_{1,k} \overset{\text{ind}}{\sim} \text{Bernoulli}(\pi_{1,k0})$ and $\tilde{\lambda}_k \overset{\text{ind}}{\sim} \text{N}^+(\mu_{1,k0}, \sigma^2_{1,k0})$, with $N^+(\mu, \sigma^2)$ denoting the $N(\mu, \sigma^2)$ distribution truncated to fall within $\Re^+$. Similarly, the priors for β_v and γ_{mk} can be induced through the specifications

$$\beta_v = (1-\delta_{2,v})\tilde{\beta}_v, \quad \delta_{2,v} \overset{\text{ind}}{\sim} \text{Bernoulli}(\pi_{2,v0}), \; \tilde{\beta}_v \overset{\text{ind}}{\sim} \text{N}(\mu_{2,v0}, \sigma^2_{2,v0}),$$

$$\gamma_{mk} = (1-\delta_{3,mk})\tilde{\gamma}_{mk}, \quad \delta_{3,mk} \overset{\text{ind}}{\sim} \text{Bernoulli}(\pi_{3,mk}), \; \tilde{\gamma}_{mk} \overset{\text{ind}}{\sim} \text{N}(\mu_{3,mk,0}, \sigma^2_{3,mk,0}).$$

In each of these cases, we simply induce a mixture prior for the coefficient through multiplying a latent indicator that the coefficient equal zero by the latent value of the coefficient if it is nonzero.

3.2 Updating Parameters

Based on the preceding settings, the SSVS algorithm alternates between steps for updating each of the unknowns as follows:

Step 1: Update $\tilde{\lambda}_k$. We first sample $\delta_{1,k}$ from its full conditional posterior distribution, Bernoulli $(\tilde{\pi}_{1,k})$, where $\tilde{\pi}_{1,k} = \frac{\pi_{1,k0}}{\pi_{1,k0}+(1-\pi_{1,k0})C_{1,k}}$,

$$C_{1,k} = \frac{\widetilde{L}(\boldsymbol{\beta}, \lambda_k = \tilde{\lambda}_k, \boldsymbol{\lambda}_{(-k)}, \boldsymbol{\gamma}, \phi; \mathbf{y})}{\widetilde{L}(\boldsymbol{\beta}, \lambda_k = 0, \boldsymbol{\lambda}_{(-k)}, \boldsymbol{\gamma}, \phi; \mathbf{y})},$$

and $\boldsymbol{\lambda}_{(-k)} = (\lambda_1, \ldots, \lambda_{k-1}, \lambda_{k+1}, \ldots, \lambda_q)'$. If $\delta_{1,k} = 1$, then let $\lambda_k = 0$ and exclude the kth random effect. Otherwise, we sample $\tilde{\lambda}_k$ for λ_k from the conditional posterior given inclusion, which is proportional to $1(\tilde{\lambda}_k > 0)$ $\widetilde{L}(\boldsymbol{\beta}, \tilde{\lambda}_k, \boldsymbol{\lambda}_{(-k)}, \boldsymbol{\gamma}, \phi; \mathbf{y})\, \text{N}(\tilde{\lambda}_k; \mu_{1,k0}, \sigma^2_{1,k0})$.

Step 2: Update $\tilde{\beta}_v$. We first sample $\delta_{2,v}$ from its full conditional posterior distribution, Bernoulli $(\tilde{\pi}_{2,v})$, where $\tilde{\pi}_{2,v} = \frac{\pi_{2,v0}}{\pi_{2,v0}+(1-\pi_{2,v0})C_{2,v}}$,

$$C_{2,v} = \frac{\widetilde{L}(\tilde{\beta}_v, \boldsymbol{\beta}_{(-v)}, \boldsymbol{\lambda}, \boldsymbol{\gamma}, \phi; \mathbf{y})}{\widetilde{L}(\tilde{\beta}_v = 0, \boldsymbol{\beta}_{(-v)}, \boldsymbol{\lambda}, \boldsymbol{\gamma}, \phi; \mathbf{y})},$$

and $\boldsymbol{\beta}_{(-v)} = (\beta_1, \ldots, \beta_{v-1}, \beta_{v+1}, \ldots, \beta_p)'$. If $\delta_{2,v} = 1$, then let $\beta_v = 0$. Otherwise, we sample $\tilde{\beta}_v$ for β_v from from the conditional posterior given inclusion, which is proportional to $\widetilde{L}(\tilde{\beta}_v, \boldsymbol{\beta}_{(-v)}, \boldsymbol{\lambda}, \boldsymbol{\gamma}, \phi; \mathbf{y})\mathrm{N}(\tilde{\beta}_v; \mu_{2,v0}, \sigma^2_{2,v0})$.

Step 3: Update $\tilde{\gamma}_{mk}(m > k)$. We first sample $\delta_{3,mk}$ from its full conditional posterior distribution, Bernoulli $(\tilde{\pi}_{3,mk})$, where $\tilde{\pi}_{3,mk} = \frac{\pi_{3,mk,0}}{\pi_{3,mk,0}+(1-\pi_{3,mk,0})C_{3,mk}}$, with $C_{3,mk} = \widetilde{L}(\boldsymbol{\beta}, \boldsymbol{\lambda}, \tilde{\gamma}_{mk}, \boldsymbol{\gamma}_{(-mk)}, \phi; \mathbf{y}) \,/ \widetilde{L}(\boldsymbol{\beta}, \boldsymbol{\lambda}, \tilde{\gamma}_{mk} = 0, \boldsymbol{\gamma}_{(-mk)}, \phi; \mathbf{y})$, where $\boldsymbol{\gamma}_{(-mk)} = (\gamma_{m'k'} : m' = k' + 1, \ldots, q; k' = 1, \ldots, k-1, k+1, \ldots, q-1)'$. If $\delta_{3,mk} = 1$, then let $\gamma_{mk} = 0$. Otherwise, we sample $\tilde{\gamma}_{mk}$ for γ_{mk} from from the conditional posterior given inclusion, which is proportional to $\widetilde{L}(\boldsymbol{\beta}, \boldsymbol{\lambda}, \tilde{\gamma}_{mk}, \boldsymbol{\gamma}_{(-mk)}, \phi; \mathbf{y})\mathrm{N}(\tilde{\gamma}_{mk}; \mu_{3,mk,0}, \sigma^2_{3,mk,0})$. However, if $\lambda_m = 0$ or $\lambda_k = 0$, $\gamma_{mk} = 0$ according to its constraint related to $\boldsymbol{\lambda}$.

Step 4: Update σ^{-2}. In the case of identity link, $\phi = \sigma^2$. We sample σ^{-2} from its full conditional distribution, $\mathcal{G}(\sigma^{-2}; c_0, d_0)\widetilde{L}(\boldsymbol{\beta}, \boldsymbol{\lambda}, \boldsymbol{\gamma}, \sigma^{-2}; \mathbf{y})$.

Samples from the joint posterior distribution of the parameters are generated by repeating these steps for a large number of iterations after apparent convergence. In general, there are no explicit forms for the full conditionals of the parameters based on the proposed approximation. However, when (10) holds, the more explicit full conditional posterior distributions for $\tilde{\lambda}_k$, $\tilde{\gamma}_{mk}$ and σ^{-2} can be derived from the joint posterior distribution as follows:

- Update $\tilde{\lambda}_k$, $k = 1, \ldots, q$ from its full conditional distribution, which is proportional to

$$1(\tilde{\lambda}_k > 0)C_{1,k}\mathrm{N}(\tilde{\lambda}_k; \mu_{1,k0}, \sigma^2_{1,k0}),$$

  where

$$C_{1,k} = \exp\left[\frac{\tilde{\lambda}_k}{2\phi}\sum_{i=1}^{n}\left\{\tilde{\lambda}_k\left(1+\sum_{s=1}^{k-1}\gamma_{ks}^2\right)B_{i,k}^{(1)} + 2\sum_{t=1,t\neq k}^{q}\lambda_t\left(\gamma_{w(t,k)} + \sum_{s=1}^{r-1}\gamma_{ks}\gamma_{ts}\right)B_{i,w(t,k)}^{(2)}\right\}\right]$$

  with $r = \min(t,k)$, $\gamma_{w(t,k)}$ equals γ_{kt} if $t < k$ and γ_{tk} otherwise, and $B^{(2)}_{i,w(t,k)}$ denotes $B^{(2)}_{i,k,t}$ if $t < k$ and $B^{(2)}_{i,t,k}$ otherwise.
- Update $\tilde{\gamma}_{mk}$ for $m > k$ from its full conditional distribution which is proportional to

$$\exp\left\{\frac{\lambda_m\tilde{\gamma}_{mk}}{\phi}\sum_{i=1}^{n}\left(\frac{1}{2}\lambda_m\tilde{\gamma}_{mk}B_{i,m}^{(1)} + \sum_{t=k,t\neq m}^{q}\lambda_t\gamma_{tk}B_{i,w(t,m)}^{(2)}\right)\right\}\mathrm{N}\left(\tilde{\gamma}_{mk}; \mu_{3,mk,0}, \sigma^2_{3,mk,0}\right).$$

- Update σ^{-2}, if $\phi = \sigma^2$, from the full conditional distribution

$$\mathcal{G}\left(c_0, d_0 - \log\left\{\widetilde{L}(\boldsymbol{\beta}, \boldsymbol{\lambda}, \boldsymbol{\gamma}, \sigma^{-2}; \mathbf{y})/L_0\right\}\right).$$

By varying the elements of $\boldsymbol{\lambda}$, $\boldsymbol{\beta}$ and $\boldsymbol{\gamma}$ that are assigned 0 values, the algorithm effectively generates samples from the posterior distribution of M. As in SSVS algorithms for linear regression, we do not visit all the possible models in $\mathcal{M}$, since this number is typically enormous. Instead, by stochastically making local changes to the model based on (approximated) conditional model probabilities, we tend to visit models with relatively high posterior probability. However, for very large model spaces, there is no guarantee that we will visit the best model in $\mathcal{M}$. In addition, there may be a large number of models which have similar posterior probability. Hence, inferences are often based on marginal posterior probabilities of excluding a particular predictor from the fixed and/or random effects components.

3.3 Calculation of Quantities

Posterior model probabilities can be estimated by averaging indicator variables across iterations collected after apparent convergence. For example, to estimate the posterior probability of the kth random effect being excluded, one can simply add up the number of iterations for which $\lambda_k = 0$ and divide by the total number of iterations. An alternative method is to use a Rao–Blackwell estimator $\widehat{\Pr}(\lambda_k = 0|\text{data}) = \frac{1}{S}\sum_{s=1}^{S}\tilde{\pi}_{1,k}^{(s)}$, where $\tilde{\pi}_{1,k}^{(s)}$ is the value of $\tilde{\pi}_{1,k}$ at iteration s, for $s = 1, \ldots, S$. This estimator is potentially more efficient. The same approach can be used to calculate posterior probabilities of excluding predictors from the fixed effect component. To estimate the posterior probability that two random effects are uncorrelated given that they are both in the model (e.g. $\sigma_{mk} = 0$), one can use the following estimator:

$$\widehat{\Pr}(\sigma_{mk} = 0|\lambda_m > 0, \lambda_k > 0, \text{data}) = \frac{\sum_{s:\lambda_m>0,\lambda_k>0} 1\left\{\rho(\zeta_{im}, \zeta_{ik}; \boldsymbol{\gamma}^{(s)}) = 0\right\}}{\sum_{s=1}^{S} 1(\lambda_m > 0, \lambda_k > 0)},$$

so that we calculate the proportion of samples for which the random effects are uncorrelated from among the samples for which both random effects are in the model.

Note that, in allowing the elements of $\boldsymbol{\Sigma}$ to have values exactly equal to zero, we obtain a shrinkage estimator of the random effects covariance. This estimator should have lower variance than typical estimators for the random effects covariance, particularly when the number of candidate random effects is moderate to large. However, the theoretical properties of shrinkage estimators of this form remain to be established.

It is important to note that typical ideas of MCMC convergence cannot be realistically applied when the MCMC algorithm is required to simultaneously search over a very high-dimensional model space, while also obtaining draws from the posterior for the parameters within each model. Simply and honestly put, when there are hundreds of thousands or even millions of models under consideration, we have no hope whatsoever of obtaining proper convergence, so that the MCMC draws can be interpreted as samples from the target distribution corresponding to the joint posterior distribution. Typical implementations of SSVS will visit a very small fraction of the models in the list if the number of models in the list is very large. Hence, we are attempting to estimate posterior model probabilities for a large number of models that have not even been visited. Nonetheless, SSVS algorithms tend to rapidly identify good models, and marginal inclusion probabilities for the predictors tend to be high for important predictors and robust across multiple chains started from different locations in $\mathcal{M}$. This warning is just to note that the Bayesian approach is very useful in this context, but does not magically provide a perfect solution to the high-dimensional model uncertainty problem.

4 Simulation Examples

4.1 Simulation Setup

As a proof of concept to assess the behavior of the approach, we carried out a simulation study. Because the SSVS approach is computationally expensive even for a single data set, it was not feasible to run a full simulation study to assess frequentist operating characteristics, such as type I error rates, power, bias in parameter estimation, and efficiency. Instead, we ran a small number of simulations under each of a variety of cases involving different random effects covariance structure from the GLMM with identity link, logit link, and log link.

We considered 100 subjects, each of which has six observations. The numbers of candidate predictors in the two components, p and q, are chosen as $p = q = 3, 5$ or 8. The covariates are $\mathbf{x}_{ij} = (x_{ij1}, \ldots, x_{ijp})'$, where $x_{ij1} = 1$ and $x_{ijk} \sim$ Bernoulli(0.5), for $i = 1, \ldots, 100$, $j = 1, \ldots, 6$, $k = 2, \ldots, p$. Let $\mathbf{z}_{ij} = \mathbf{x}_{ij}$, $\boldsymbol{\beta}_{(-2)} \sim \mathrm{N}(0, \mathbf{I})$, $\beta_2 = 0$, and $\boldsymbol{\zeta}_i = (\zeta_{i1}, \ldots, \zeta_{iq})' \sim \mathrm{N}(\mathbf{0}, \boldsymbol{\Sigma})$, where we designed $\boldsymbol{\Sigma} = \boldsymbol{\Lambda\Gamma\Gamma'\Lambda}$ with three different structures:

(1) $\boldsymbol{\lambda} = (1.2, 0.4, 0.6)'$ and $\boldsymbol{\gamma} = (0.4, 0.5, 0.3)'$, implying that all three random effects are included in the model.
(2) $\boldsymbol{\lambda} = (0.2, 0, 0.7, 0, 0.5)'$ and $\boldsymbol{\gamma} = (0, 0.4, 0, 0, 0, 0, 0.8, 0, 0.1, 0)'$, implying that the second and the fourth random effects are excluded from the model.
(3) $\boldsymbol{\lambda} = (0.5, 0.8, 0.9, 0.2, 0.1, 0.1, 0.6, 0)'$ and $\boldsymbol{\gamma} = (0.3, 0.6, 0.5, 0.4, 0.2, 0.1, 0.2, 0.3, 0.4, 0.3, 0.6, 0.1, 0.2, 0.1, 0.8, 0.3, 0.4, 0.8, 0.6, 0.3, 0.2, 0, 0, 0, 0, 0, 0, 0)'$, implying that the last random effect is excluded from the model.

The corresponding covariance matrices for random effects are shown in the first row in Fig. 1. For the GLMM with identity link, $y_{ij} \sim \mathrm{N}(\mathbf{x}_{ij}'\boldsymbol{\beta} + \mathbf{z}_{ij}'\boldsymbol{\zeta}_i, \sigma^{-2})$ with $\sigma^{-2} = 2$. For the GLMM with logit link, $y_{ij} \sim \mathrm{Bernoulli}(\pi_{ij})$ with $\mathrm{logit}(\pi_{ij}) = \mathbf{x}_{ij}'\boldsymbol{\beta} + \mathbf{z}_{ij}'\boldsymbol{\zeta}_i$. For the GLMM with log link, $y_{ij} \sim \mathrm{Poisson}(\lambda_{ij})$ with $\log(\lambda_{ij}) = \mathbf{x}_{ij}'\boldsymbol{\beta} + \mathbf{z}_{ij}'\boldsymbol{\zeta}_i$.

We chose the prior distribution for λ_k as ZI-N$^+(\lambda_k; \pi_{1,k0}, 0, 10)$. The prior distributions for the elements of $\boldsymbol{\gamma}$ are chosen to be mixture priors, ZI-N$(\gamma_{mk}; \pi_{3,u0}, 0, 1)$,

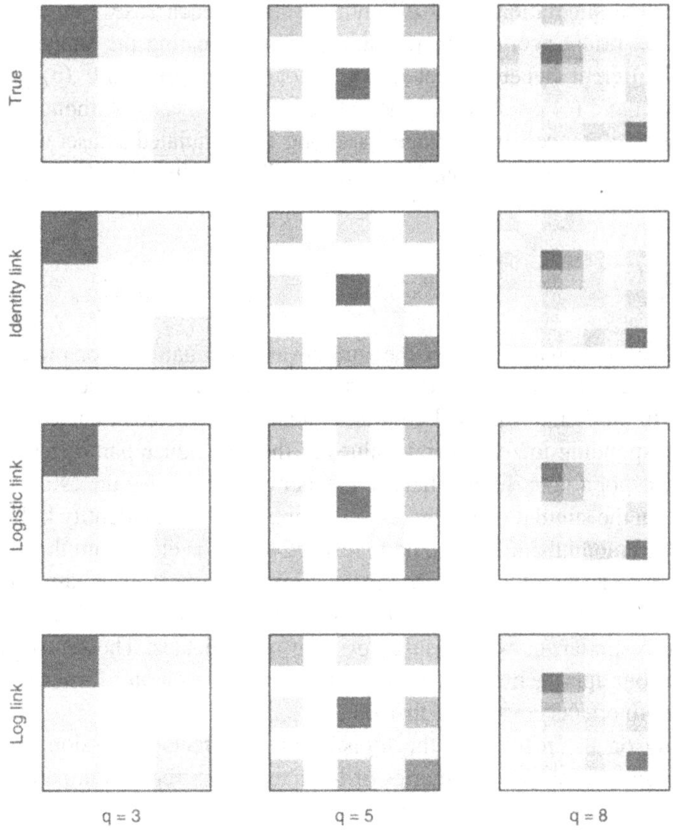

Fig. 1 Image plots of the true and estimated random effects covariance matrices for simulated data under the identity link, logit link and log link with the number of candidate random effect predictors being 3, 5 and 8. The darker the color appears, the larger the value of the element is, with the white color corresponding to zero

with the constraint related to $\boldsymbol{\lambda}$. A mixture prior distribution for β_v is chosen as ZI-N$(\beta_v; \pi_{2,k0}, 0, 10)$. A diffuse prior for parameter σ^{-2} is chosen to be $\mathcal{G}(0.08, 0.08)$. To study the effects of the prior probabilities of $\lambda_k = 0$, $\beta_v = 0$ and $\gamma_{mk} = 0$ on the estimated posterior probabilities, we consider 0.2, 0.5, 0.8 for these prior probabilities.

For each simulated data set and choice of prior, we ran the Gibbs sampling algorithm described in Sect. 3 for 20,000 iterations after a 2,000 burn-in. The diagnostic tests were carried out by using Geweke (1992) and Raftery and Lewis (1992), which showed rapid convergence and efficient mixing. Note that this apparent good performance in terms of convergence and mixing is somewhat counter to our warning at the end of Sect. 3. A sample of size 4,000 was obtained by thinning the MCMC chain by a factor of 5. For each simulated data set, we calculated (a) the posterior probabilities for the possible submodels under each link; and (b) the estimated posterior means and the 95% credible intervals for each of the parameters.

We ran five simulations with different seeds for each case. Sensitivity of the results to the prior specification was assessed by repeating the analyses with the following different hyperparameters: (a) priors with variance/2; (b) priors with variance × 2; and (c) priors with moderately different means. Although we do not show details, inferences for all models are robust to simulated dataset with different seeds and the prior specification. The ranges in Table 2 illustrate this robustness.

4.2 Results

Figure 1 displays image plots of the true covariance matrices for random effects corresponding to the simulated data and the estimated covariance matrices under each link. In particular, the final column contains the true covariance matrix, with white corresponding to zero or low values of the covariance parameter and darker shades corresponding to high values. The other columns show the estimated posterior means in the simulation examples under log, logistic and identity link functions for Poisson, binomial and normal data, respectively. It is clear from these plots that we obtained an accurate estimate of the covariance matrix in each case. In addition, the estimates from the sensitivity analyses that varied the prior inclusion probabilities were similar, as were results for each of replicates. These results provide support for our approach as a method for obtained an accurate shrinkage estimator of the random effects covariance matrix.

Focusing on the results for the logistic mixed effects regression simulations, Table 1 presents posterior summaries of the fixed effect regression parameters and

Table 1 Posterior estimates of the parameters for the second simulation with the logit link

Parameter	True value	Mean	SD	95% HPD interval
β_1	−0.02	0.010	0.120	(−0.231, 0.273)
β_2	0	−0.001	0.025	(−0.002, 0.001)
β_3	−0.60	−0.611	0.166	(−0.934, −0.284)
β_4	1.48	1.448	0.142	(1.170, 1.725)
β_5	−0.81	−0.792	0.167	(−1.120, −0.467)
σ_{11}	0.04	0.037	0.011	(0.013, 0.059)
σ_{22}	0	0.007	0.021	(0, 0)
σ_{33}	0.57	0.578	0.120	(0.348, 0.820)
σ_{44}	0	0.002	0.024	(0, 0.001)
σ_{55}	0.41	0.396	0.097	(0.208, 0.587)
σ_{21}	0	0.006	0.019	(0, 0)
σ_{31}	0.06	0.052	0.016	(0.020, 0.084)
σ_{32}	0	0.002	0.026	(0, 0)
σ_{41}	0	0.001	0.030	(0, 0.001)
σ_{42}	0	0.003	0.022	(0, 0)
σ_{43}	0	0.002	0.027	(0, 0)
σ_{51}	0.08	0.073	0.028	(0.023, 0.122)
σ_{52}	0	0.000	0.024	(0, 0)
σ_{53}	0.15	0.159	0.035	(0.084, 0.223)
σ_{54}	0	0.001	0.033	(0, 0)

Table 2 Estimated model posterior probabilities in simulation studies under the logit link. Submodels with posterior probability less than 0.02 are not displayed

Model	$\pi_{1,k0}$		
	0.2	0.5	0.8
Simulation 1			
x_1, x_3, z_1, z_2, z_3 [a]	$0.833^{b}_{(0.814,0.865)^{c}}$	$0.796_{(0.771,0.828)}$	$0.748_{(0.719,0.782)}$
x_1, x_3, z_1, z_3	$0.085_{(0.054,0.116)}$	$0.098_{(0.083,0.115)}$	$0.116_{(0.098,0.141)}$
x_1, x_3, z_1, z_2	$0.066_{(0.045,0.092)}$	$0.070_{(0.046,0.095)}$	$0.082_{(0.059,0.106)}$
Simulation 2			
$x_1, x_3, x_4, x_5, z_1, z_3, z_5$ [a]	$0.437_{(0.421,0.544)}$	$0.519_{(0.483,0.558)}$	$0.568_{(0.543,0.591)}$
$x_3, x_4, x_5, z_1, \ldots, z_5$	$0.106_{(0.075,0.140)}$	$0.095_{(0.068,0.131)}$	$0.084_{(0.048,0.112)}$
$x_1, x_3, x_4, x_5, z_3, z_5$	$0.103_{(0.076,0.135)}$	$0.139_{(0.110,0.180)}$	$0.177_{(0.138,0.218)}$
$x_3, x_4, x_5, z_1, z_3, z_4, z_5$	$0.024_{(0.013,0.037)}$	$0.039_{(0.026,0.054)}$	$0.052_{(0.037,0.078)}$
$x_2, \ldots, x_5, z_1, \ldots, z_4$	$0.022_{(0.010,0.035)}$	$0.037_{(0.020,0.053)}$	$0.045_{(0.024,0.066)}$
Simulation 3			
$x_1, x_3, \ldots, x_8, z_1, \ldots, z_7$ [a]	$0.547_{(0.529,0.577)}$	$0.581_{(0.550,0.617)}$	$0.633_{(0.602,0.658)}$
$x_1, x_3, \ldots, x_8, z_1, \ldots, z_8$	$0.090_{(0.079,0.106)}$	$0.075_{(0.065,0.089)}$	$0.064_{(0.053,0.075)}$
$x_1, x_3, \ldots, x_7, z_1, \ldots, z_4, z_6, z_7$	$0.051_{(0.043,0.059)}$	$0.074_{(0.066,0.085)}$	$0.078_{(0.070,0.089)}$
$x_1, x_3, \ldots, x_7, z_1, \ldots, z_5, z_7$	$0.032_{(0.027,0.041)}$	$0.034_{(0.024,0.042)}$	$0.037_{(0.030,0.041)}$
$x_1, x_3, \ldots, x_7, z_1, \ldots, z_4, z_7$	$0.025_{(0.013,0.039)}$	$0.027_{(0.018,0.037)}$	$0.032_{(0.023,0.044)}$
$x_1, \ldots, x_7, z_1, \ldots, z_3, z_7$	$0.023_{(0.011,0.033)}$	$0.025_{(0.014,0.035)}$	$0.028_{(0.016,0.040)}$

[a] True model
[b] Posterior probability
[c] Range

random effects covariance parameters in the second simulation case. It is clear that the posterior mean values are very close to the true values in each case, and that the true values are included in 95% HPD intervals. Again, we obtained similar results in sensitivity analyses and other simulation cases.

Table 2 shows the estimated posterior model probabilities for the preferred models under the three different simulation cases for a range of values for the prior probability of excluding a predictor. Although the estimated posterior probabilities varied somewhat as the prior exclusion probabilities varied, the rankings in the models were robust. In each case, the true model was the dominate model, having substantially higher estimated posterior probability than the second best model. The ranges shown in subscripts represent the range in the estimated posterior model probabilities across the five different simulation replicates and for different choices of hyperparameters. Figure 2 presents boxplots of the samples of parameters for the second simulation under each link. The true values of all parameters fall in the 95% credible intervals.

When the number of models under consideration is large, the posterior probability assigned to any one model is typically not close to 1. This property is observed in Table 2. Although the highest posterior probability is assigned to the correct model

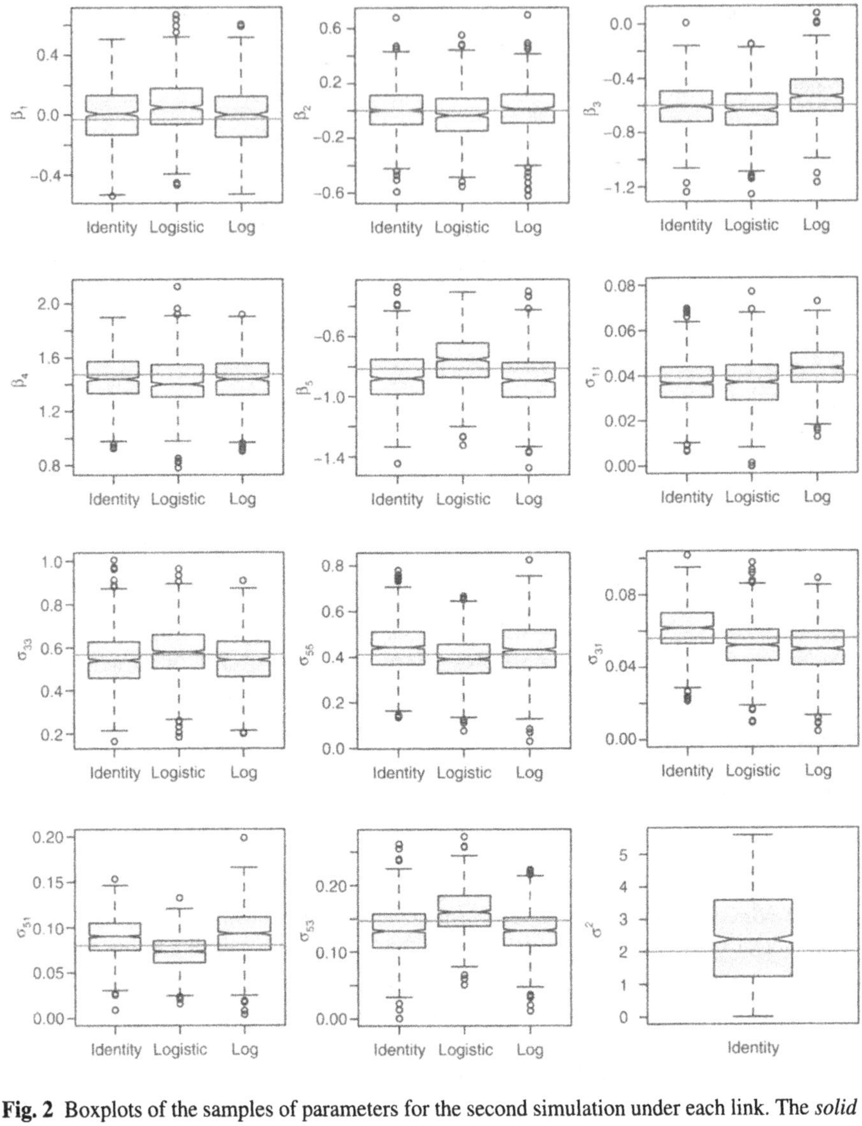

Fig. 2 Boxplots of the samples of parameters for the second simulation under each link. The *solid horizontal lines* indicate the true values

in each case, this probability is never close to one. For example, in simulation case 3, the probability assigned to the true model is only slightly above 0.5. This does not suggest that our approach for estimating posterior probabilities is poor, but is instead a general feature of model selection in large model spaces.

For this reason, in performing inferences on whether a given predictor should be included in the fixed and/or random effects component, it is more reliable to rely on marginal inclusion probabilities than on whether that predictor is included in

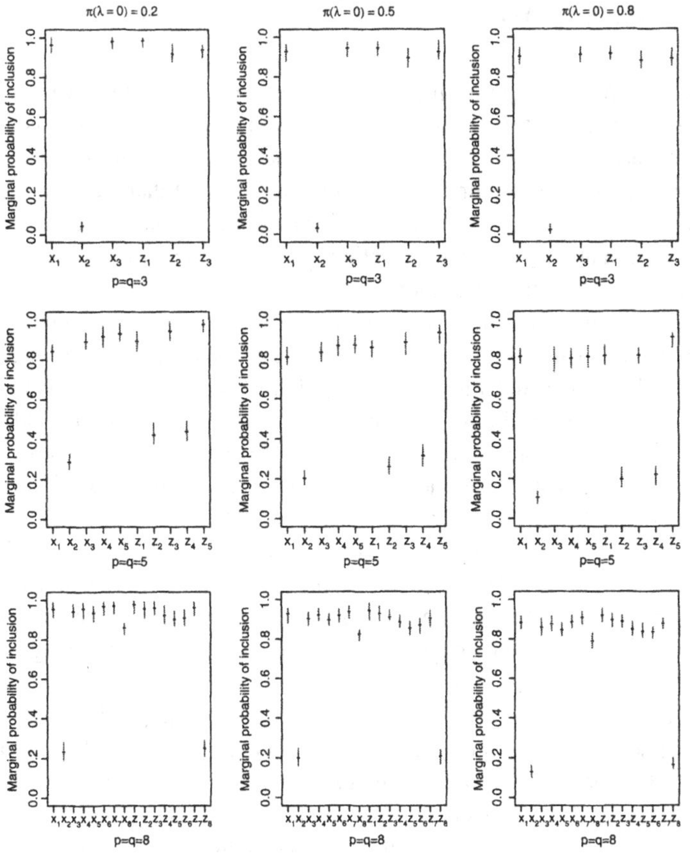

Fig. 3 Plots of marginal inclusion probabilities for each of the predictors in terms of fixed and random effects under the logistic model with different prior probabilities (0.2, 0.5, and 0.8) of exclusion of each predictor and the different number (3, 5, and 8) of candidate predictors. The short *horizontal lines* denote the posterior marginal inclusion probabilities for each of the predictors. The *vertical lines* show the ranges of marginal inclusion probabilities on average

the highest posterior probability model. Figure 3 shows plots of posterior marginal inclusion probabilities of each of the predictors in terms of fixed and random effects under the logistic model with different prior probabilities (0.2, 0.5, and 0.8) of exclusion of each predictor and the different number (3, 5, and 8) of candidate predictors. It is clear that although marginal inclusion probabilities for each of the predictors change slightly according to the different choices of prior probabilities of exclusion, the inclusion of the predictors is consistent with the designed models. For example, the third row presents the marginal inclusion probabilities of eight fixed and random effects predictors with different prior probabilities of exclusion (0.2, 0.5, and 0.8). Obviously, the predictors for the second fixed effect and for the eighth random effect are less likely to be included in the model since the corresponding marginal inclusion probabilities are fairly low. We also calculated marginal inclusion probabilities of predictors for the GLMM with log link and identity link, which show the consistent results with the true models.

4.3 Assessment of Accuracy of the Approximation

Although the proposed approach appears to perform well at model selection and parameter estimation based on the results in the simulation examples, it is important to assess the accuracy of the proposed approximation to the marginal likelihood. If the approximation is not accurate, then it may be the case that our approach is not producing accurate estimates of the posterior model probabilities. Even if we do a good job in estimation and model selection, it does not necessarily imply accuracy in marginal likelihood approximation.

Because it tends to be the case that Taylor series-based approximations to the marginal likelihood are less accurate for binary response logistic mixed effects models than for log-linear and linear cases, we focus on the logistic special case in assessing accuracy. In addition, we focus on the modest sample size of 100 subjects, each of which has six observations, and we let $p = q = 3$. We choose different covariance structures with variance components proportional to $\boldsymbol{\lambda}^2$ from small to large, which are (1) $(0.01, 0.02, 0.005)'$; (2) $(1.2, 0.4, 0.6)'$; (3) $(2.8, 4.3, 3.5)'$; (4) $(27.5, 20.6, 35.1)'$; (5) $(50.6, 30.8, 60.3)'$, with $\boldsymbol{\gamma}$ kept fixed at $(0.4, 0.5, 0.3)'$.

An essentially exact value for the marginal likelihood can be obtained by brute force numerical integration, so we use that approach as the reference in comparing several approximation approaches. In particular, we estimate the marginal likelihoods of the simulated data using the Laplace approximation, importance sampling, Chib's marginal likelihood method, and our proposed approach. Table 3 shows the comparison of log marginal likelihoods calculated by the different methods. Note that all of the approximations tend to perform better when the random effects variance is small, with the accuracy decreasing for large random effects variances. For small variances, the proposed approach was slightly more accurate than any of the competitors. In addition, the proposed approach was closer to the truth than the Laplace approximation in all the cases.

Note that the approximation to the marginal likelihood is only used in calculating the conditional posterior probabilities that a coefficient is equal to zero. Hence, when there is clear evidence that the predictor should be included, some inaccuracy in the marginal likelihood approximation has no impact on inferences. Therefore, under our approach the performance for small values of the random effect variances is most important. However, given the improvement seen relatively to the widely-used Laplace approximation for all values of the random effects variance, the proposed approximation should also be useful in frequentist inferences and other settings.

Table 3 Comparison of approximated log marginal likelihoods for the GLMM with the logit link

$\boldsymbol{\lambda}^2$	Chib's	Exact	I.Sampling	Laplace	Proposed
$(0.01, 0.02, 0.005)$	−92.47	−89.19	−92.09	−91.74	−91.59
$(1.2, 0.4, 0.6)$	−129.23	−125.35	−128.68	−131.05	−128.15
$(2.8, 4.3, 3.5)$	−147.37	−140.80	−145.01	−148.39	−148.22
$(27.5, 20.6, 35.1)$	−94.83	−88.93	−93.62	−97.92	−96.47
$(50.6, 30.8, 60.3)$	−100.56	−90.97	−98.81	−104.73	−101.86

5 Time-to-Pregnancy Application

5.1 Data and Model Selection Problem

Returning to the time to pregnancy (TTP) application introduced in Sect. 1.2, we analyze the Rowland et al. (1992) study to illustrate the proposed approach. This was a retrospective TTP study, with female dental assistants, aged 19–39, contacted and invited to enrolled after being randomly selected from a registry. Study investigators enrolled 427 women, who completed a detailed questionnaire on reproductive history, occupational exposures and other factors that may be related to fecundability. As illustration, we focus on the logistic mixed effects discrete hazard model presented in (1), with the candidate predictors including category indicators for age (19–24, 25–29, >30), intercourse frequency per week (≤ 1, 1–3, 3–4, >4), cigarettes smoked per day (nonsmoker, 1–5, 6–10, 11–15, >15), and the use of oral contraceptives in the cycle prior to beginning the pregnancy attempt (no, yes).

Including all the above indicator variables in the order that they are introduced, with the first category level being the reference, we have 14 candidate predictors. We allow each of these predictors to be potentially included in the fixed and/or random effects component. Hence, we have an enormous list of possible models when also allowing uncertainty in whether the random effects correlations are equal to zero.

5.2 Prior Specification, Implementation and Results

In choosing a prior, our goal was to assign high probability to a wide range of plausible values for the regression coefficients and induced covariance matrix without specifying a very high variance prior. Given that all the predictors are indicator variables within a logistic regression model, a prior variance of 20 (given inclusion in the model) for both λ_k and β_υ seemed reasonable. Hence, the prior distribution for λ_k is chosen as ZI-N$^+(\lambda_k|\pi_{k0}, 0, 20)$, and the prior distribution for β_υ is chosen as ZI-N$(\beta_\upsilon|\pi_{\upsilon 0}, 0, 20)$. For the elements of $\boldsymbol{\gamma}$, we choose a more informative prior with a variance of one to favor modest levels of correlation by letting ZI-N$(\gamma_{mk}; \pi_{3,u0}, 0, 1)$, with the constraint on $\boldsymbol{\lambda}$.

In considering applications with continuous predictors, one should either normalize these predictors prior to analysis, or carefully consider the measurement scale of the predictors in choosing the prior variance. How informative the prior is for a particular prior variance is completely dependent on the scale of the measurements. Note that the tendency is to favor smaller models the larger the prior variance. Our choice of 20 is really in the upper range of reasonable values for the prior variance in this context, and our motivation was to favor parsimony.

We ran the MCMC algorithm for 80,000 iterations after a 10,000 iteration burn-in. This chain was a bit longer than is typical for posterior computation in a single GLMM. In general, when the MCMC algorithm is being used to simultaneously explore a high-dimensional model space, while also estimating posterior model probabilities and posterior distributions of coefficients, the MCMC chain should be dramatically longer than that used in analysis of a single model. The chains passed Geweke (1992) and Raftery and Lewis (1992) convergence diagnostic tests. Multiple chains with different initial settings passed Gelman and Rubin (1992) convergence test. We retained every 20th sample for inferences of interest.

As we discussed in Sect. 4.2, when the number of models in the list is enormous, the posterior probability assigned to any one model tends to be quite small, even if that model is the true model. This behavior is expected and provides strong support for Bayesian methods of model averaging and inferences, which avoid relying on interpreting any single selected model as being supported by the data. Frequentist or Bayesian methods for selecting an optimal model tend to ignore the issue that it is effectively impossible to find clear evidence in favor of the true model in a very large list unless the amount of data you have available is incredibly massive. Hence, it is much more reasonable to focus on marginal inclusion probabilities in gauging importance of predictors when the list of possible models is huge. In the TTP application, the top models had estimated posterior probabilities close to 0.03, with several candidates having similar values.

Table 4 presents the marginal posterior probabilities of including each predictor in the fixed and random effects components under different choices of $\pi_{1,k0}$ and $\pi_{2,\upsilon 0}$. The overall posterior probability of including age in the fixed effects component can be calculated as the posterior probability that any of the category indicators for age are included. Such overall posterior probabilities are calculated separately for the fixed and random effects components for each of the factors under consideration, including age, intercourse frequency, cigarettes smoked, and recent pill use. The results are shown in Table 4. The posterior probability of including age in the fixed effects component ranges from 0.95 to 0.97 (average = 0.96) depending on the prior. The corresponding ranges for intercourse frequency, cigarettes smoked, and recent pill use are 0.96–0.97 (average = 0.96), 0.92–0.97 (average = 0.94), and 0.87–0.98 (average = 0.92), respectively. Hence, as expected, there is some evidence that age, intercourse frequency, cigarette smoking, and recent pill use are predictive of fecundability on average, with the most evidence for age and intercourse frequency. The age effect is most apparent in women 30+. In addition, the indicators for the highest categories of intercourse frequency (4+ acts/week) and cigarette smoking (15+/day) had the highest posterior probabilities of inclusion.

For the random effects component, the results were somewhat different. The posterior probability of inclusion for recent pill use ranged between 0.40 and 0.53 (average = 0.47), so there is no evidence of heterogeneity in the effect of recent pill use. However, there was some evidence of heterogeneity among women in the effects of each of the other factors. In particular, the posterior probability of including age in the random effects component ranged between 0.87 and 0.92 (average = 0.90), which is suggestive but not clear evidence. There was slightly more evidence of heterogeneity in the effects of intercourse frequency and cigarette smoking with the posterior probabilities of inclusion for these two factors ranging between 0.90 and 0.93 (average = 0.92) and 0.91–0.95 (average = 0.93), respectively.

Table 4 Estimated marginal posterior probabilities of including predictors in the fixed and random effects components under different prior probabilities of being zero in the time-to-pregnancy application. Probabilities over 0.9 are written in bold

Predictor	Posterior probability of inclusion					
	Fixed effect			Random effect		
	0.2	0.5	0.8	0.2	0.5	0.8
Intercept	**0.90**(0.89,0.94)[a]	0.87(0.85,0.90)	0.83(0.81,0.87)	**0.94**(0.90,0.96)	**0.90**(0.86,0.92)	0.85(0.82,0.87)
Age						
25–29	0.83(0.80,0.85)	0.75(0.73,0.78)	0.72(0.69,0.75)	0.55(0.52,0.58)	0.50(0.46,0.52)	0.43(0.39,0.46)
30+	**0.93**(0.90,0.96)	0.89(0.86,0.92)	0.87(0.84,0.90)	0.88(0.84,0.91)	0.86(0.83,0.90)	0.81(0.76,0.84)
Overall	**0.97**(0.94,0.99)	**0.96**(0.94,0.98)	**0.95**(0.93,0.97)	**0.92**(0.89,0.95)	**0.90**(0.87,0.92)	0.87(0.84,0.91)
Intercourse frequency						
1–3	0.63(0.61,0.66)	0.56(0.54,0.59)	0.53(0.49,0.56)	0.61(0.56,0.64)	0.56(0.52,0.59)	0.52(0.49,0.56)
3–4	0.76(0.73,0.79)	0.73(0.71,0.76)	0.68(0.65,0.71)	0.83(0.78,0.87)	0.78(0.75,0.82)	0.74(0.70,0.78)
4+	**0.97**(0.93,0.98)	**0.94**(0.91,0.96)	0.88(0.85,0.91)	0.54(0.49,0.57)	0.50(0.45,0.53)	0.44(0.40,0.48)
Overall	**0.97**(0.94,0.99)	**0.96**(0.94,0.98)	**0.96**(0.93,0.98)	**0.93**(0.90,0.96)	**0.92**(0.89,0.94)	**0.90**(0.86,0.93)
Cigarettes smoked						
1–5	0.70(0.68,0.73)	0.65(0.62,0.67)	0.56(0.52,0.59)	0.61(0.59,0.65)	0.55(0.52,0.59)	0.51(0.47,0.53)
6–10	0.85(0.81,0.88)	0.81(0.77,0.83)	0.74(0.72,0.78)	0.89(0.85,0.92)	0.85(0.82,0.88)	0.82(0.77,0.85)
11–15	0.86(0.82,0.88)	0.79(0.76,0.82)	0.70(0.67,0.74)	0.57(0.53,0.61)	0.52(0.49,0.55)	0.49(0.46,0.54)
15+	**0.95**(0.92,0.97)	**0.92**(0.89,0.94)	0.88(0.84,0.91)	0.69(0.65,0.75)	0.66(0.62,0.70)	0.61(0.58,0.65)
Overall	**0.97**(0.94,0.99)	**0.94**(0.91,0.98)	**0.92**(0.89,0.95)	**0.95**(0.93,0.98)	**0.92**(0.89,0.95)	**0.91**(0.87,0.94)
Recent pill use	**0.98**(0.96,0.99)	**0.91**(0.87,0.95)	0.87(0.85,0.91)	0.53(0.48,0.59)	0.48(0.43,0.51)	0.40(0.37,0.45)

[a]Range

Table 5 provides the overall posterior summaries of the regression coefficients from our approach compared with the standard GLM with the logit link fitted to the full model. It is clear that there are no systematic differences between our model-averaged Bayesian point and interval estimates for the regression coefficients and the maximum likelihood estimates. In general, the results can differ substantially, particularly when the focus is on inferences, and the frequentist analyst selects the final model based on typical criteria (e.g., using stepwise selection), while the Bayesian uses model averaging. The results in this case, however, are due to the fact that most of the candidate predictors have moderate to high probabilities of being included in the model. We also note that the coefficient for the oldest level of the age variable is larger than for the younger levels, implying that the probability of getting pregnant goes up as age goes up which seems counter intuitive. The counter-intuitive age effect was also apparent in frequentist and more routine Bayesian analyses. One of the difficulties in time to pregnancy studies is that it is

Table 5 Overall posterior means and 95% credible intervals of regression coefficients in the time-to-pregnancy application compared with the results from the GLM with the logit link

Effects	Proposed approach	Standard GLM
Age		
25–29	$0.173_{(-0.043,0.396)}$[a]	$0.177_{(-0.068,0.422)}$[b]
30+	$0.358_{(0.003,0.727)}$	$0.354_{(-0.029,0.719)}$
Intercourse frequency per week		
1–3	$0.095_{(-0.144,0.335)}$	$0.088_{(-0.151,0.327)}$
3–4	$0.258_{(-0.120,0.643)}$	$0.265_{(-0.129,0.659)}$
4+	$0.877_{(0.402,1.306)}$	$0.885_{(0.408,1.361)}$
Cigarettes smoked per day		
1–5	$-0.119_{(-0.570,0.313)}$	$-0.126_{(-0.736,0.482)}$
6–10	$-0.278_{(-0.694,0.241)}$	$-0.285_{(-0.873,0.303)}$
11–15	$-0.425_{(-1.239,0.337)}$	$-0.433_{(-1.413,0.547)}$
15+	$-0.686_{(-1.640,-0.164)}$	$-0.681_{(-1.622,0.260)}$
Use of oral contraceptives	$-0.923_{(-1.544,-0.268)}$	$-0.931_{(-1.619,-0.243)}$

[a]95% credible interval
[b]95% confidence interval

impossible to get a group of women of different ages who are at risk of pregnancy and representative of the general population of reproductive age women, particularly in an occupational epidemiology setting. It may be the case that the older dental assistants are representative of a different demographic group having higher fertility, or that some selection.

We ran extensive sensitivity analyses to evaluate the robustness of the results to the prior specifications by repeating the analyses with the following different hyperparameters: (a) priors with half variance; (b) priors with double variance; (c) priors with moderately different means within the range of the prior expectation. The ranges in Table 4 show the results for all of the different priors.

6 Discussion

This chapter has proposed a Bayesian approach for accounting for uncertainty in selection of fixed and random effects in GLMMs. The approach is very computationally intensive in relying on MCMC to simultaneously explore a very high-dimensional model space and estimate posterior model probabilities and densities for selected predictors. However, given the great deal of time and expense involved in collecting the data, it seems that spending a bit of time in implementing an improved analysis is well worth the effort. A C program is available from the authors upon request.

On the topic of Bayesian methods for GLMM variable selection, there are several areas of substantial interest for future research. The first is default prior selection. It is appealing to have general software available that produces results with good frequentist and Bayesian properties without need for careful thought in prior choice or sensitivity to subjectively chosen hyperparameters. There has been some work done on default prior selection for fitting of a single GLMM, but default prior selection in model uncertainty contexts is a completely different issue. For linear regression subset selection, mixtures of g-priors provide a useful default, but there is a lack of similar priors for GLMMs. In the absence of carefully-justified default priors, one can use the priors proposed in this chapter after normalizing any continuous predictors.

Another important area is the development of simpler and more efficient computational implementations, particularly for cases involving massive numbers of candidate predictors. For example, it may be the case that the proposed approximation to the marginal likelihood can be expanded to marginalize out not only the random effects, but also all the parameters in a particular GLMM. This will certainly result in much more efficient computation, and may be quite appealing if the proposed approximation can be justified as accurate.

References

Albert, J.H. and Chib, S. (1997). Bayesian Test and Model Diagnostics in Conditionally Independent Hierarchical Models. *Journal of the American Statistical Association* **92**, 916–925

Breslow, N.E. and Clayton, D.G. (1993). Approximate Inference in Generalized Linear Mixed Models. *Journal of the American Statistical Association* **88**, 9–25

Cai, B. and Dunson, D.B. (2006). Bayesian Covariance Selection in Generalized Linear Mixed Models. *Biometrics* **62**, 446–457

Cai, B. and Dunson, D.B. (2007). Bayesian Variable Selection in Nonparametric Random Effects Models. *Biometrika*, under revision

Chen, Z. and Dunson, D.B. (2003). Random Effects Selection in Linear Mixed Models. *Biometrics* **59**, 762–769

Chen, M., Ibrahim, J.G., Shao, Q., and Weiss, R.E. (2003). Prior Elicitation for Model Selection and Estimation in Generalized Linear Mixed Models. *Journal of Statistical Planning and Inference* **111**, 57–76

Chipman, H., George, E.I., and McCulloch, R.E. (2002). Bayesian Treed Models. *Machine Learning* **48**, 299–320

Chipman, H., George, E.I., and McCulloch, R.E. (2003). Bayesian Treed Generalized Linear Models. *Bayesian Statistics 7* (J.M. Bernardo, M.J. Bayarri, J.O. Berger, A.P. Dawid, D. Heckerman, A.F.M. Smith, and M. West, eds), Oxford: Oxford University Press, 323–349

Daniels, M.J. and Kass, R.E. (1999). Nonconjugate Bayesian Estimation of Covariance Matrices and Its Use in Hierarchical Models. *Journal of the American Statistical Association* **94**, 1254–1263

Daniels, M.J. and Pourahmadi, M. (2002). Bayesian Analysis of Covariance Matrices and Dynamic Models for Longitudinal Data. *Biometrika* **89**, 553–566

Daniels, M.J. and Zhao, Y.D. (2003). Modelling the Random Effects Covariance Matrix in Longitudinal Data. *Statistics in Medicine* **22**, 1631–1647

Gelman, A. and Rubin, D.B. (1992). Inference from Iterative Simulation using Multiple Sequences. *Statistical Science* **7**, 457–472

George, E.I. and McCulloch, R.E. (1993). Variable Selection via Gibbs Sampling. *Journal of the American Statistical Association* **88**, 881–889

Geweke, J. (1992). Evaluating the Accuracy of Sampling-Based Approaches to the Calculation of Posterior Moments. *Bayesian Statistics 4* (J.M. Bernardo, J.O. Berger, A.P. Dawid, and A.F.M. Smith, eds), Oxford: Oxford University Press, 169–193

Geweke, J. (1996). Variable Selection and Model Comparison in Regression. *Bayesian Statistics 5* (J.O. Berger, J.M. Bernardo, A.P. Dawid, and A.F.M. Smith, eds), Oxford: Oxford University Press, 609–620

Gilks, W.R., Best, N.G., and Tan, K.K.C. (1995). Adaptive Rejection Metropolis Sampling within Gibbs Sampling. *Applied Statistics* **44**, 455–472

Gilks, W.R., Neal, R.M., Best, N.G., and Tan, K.K.C. (1997). Corrigendum: Adaptive Rejection Metropolis Sampling. *Applied Statistics* **46**, 541–542

Kinney, S.K. and Dunson, D.B. (2007). Fixed and random effects selection in linear and logistic models. *Biometrics* **63**, 690–698

Laird, N.M. and Ware, J.H. (1982). Random Effects Models for Longitudinal Data. *Biometrics* **38**, 963–974

Liechty, J.C., Liechty, M.W., and Muller, P. (2004). Bayesian Correlation Estimation. *Biometrika* **91**, 1–14

Lin, X. (1997). Variance component testing in generalised linear models with random effects. *Biometrika* **84**, 309–326

McCulloch, C.E. (1997). Maximum Likelihood Algorithms for Generalized Linear Mixed Models. *Journal of the American Statistical Association* **92**, 162–170

McCulloch, C.E. and Searle, S. (2001). Generalized Linear and Mixed Models. New York: Wiley.

McGilchrist, C.A. (1994). Estimation in Generalized Mixed Models. *Journal of the Royal Statistical Society B* **56**, 61–69

Meyer, M.C. and Laud, P.W. (2002). Predictive Variable Selection in Generalized Linear Models. *Journal of the American Statistical Association* **97**, 859–871

Nott, D.J. and Leonte, D. (2004). Sampling Schemes for Bayesian Variable Selection in Generalized Linear Models. *Journal of Computional and Graphical Statistics* **13**, 362–382

Ntzoufras, I., Dellaportas, P., and Forster, J.J. (2003). Bayesian Variable Selection and Link Determination for Generalised Linear Models. *Journal of Statistical Planning and Inference* **111**, 165–180

Raftery, A. (1996). Approximate Bayes Factors and Accounting for Model Uncertainty in Generalized Linear Models. *Biometrika* **83**, 251–266

Raftery, A.E. and Lewis, S. (1992). How Many Iterations in the Gibbs Sampler? *Bayesian Statistics 4* (J.M. Bernardo, J.O. Berger, A.P. Dawid, and A.F.M. Smith, eds), Oxford: Oxford University Press, 763–773

Raftery, A.E., Madigan, D., and Volinsky, C.T. (1996). Accounting for Model Uncertainty in Survival Analysis Improves Predictive Performance. *Bayesian Statistics 5* (J.M. Bernardo, J.O. Berger, A.P. Dawid, and A.F.M. Smith, eds), Oxford: Oxford University Press, 323–349

Rowland, A.S., Baird, D.D., Weinberg, C.R., Shore, D.L., Shy, C.M., and Wilcox, A.J. (1992). Reduced Fertility Among Women Employed as Dental Assistants Exposed to High Levels of Nitrous Oxide. *The New England Journal of Medicine* **327**, 993–997

Schall, R. (1991). Estimation in Generalized Linear Mixed Models with Random Effects. *Biometrika* **78**, 719–727

Sinharay, S. and Stern, H.S. (2001). Bayes Factors for Variance Component Testing in Generalized Linear Mixed Models. In *Bayesian Methods with Applications to Science, Policy and Official Statistics (ISBA 2000 Proceedings)*, 507–516

Solomon, P.J. and Cox, D.R. (1992). Nonlinear Component of Variance Models. *Biometrika* **79**, 1–11

Wong, F., Carter, C.K., and Kohn, R. (2003). Efficient Estimation of Covariance Selection Models. *Biometrika* **90**, 809–830

Zeger, S.L. and Karim, M.R. (1991). Generalized Linear Models with Randome Effects: A Gibbs Sampling Approach. *Journal of the American Statistical Association* **86**, 79–86

Appendix

The normal linear mixed model of Laird and Ware (1982) is a special case of a GLMM having $g(\mu_{ij}) = \mu_{ij} = \eta_{ij} = \mathbf{x}'_{ij}\boldsymbol{\beta} + \mathbf{z}'_{ij}\boldsymbol{\zeta}_i$, $\phi = \sigma^2$ and $b(\theta_{ij}) = \eta^2_{ij}/2$. In this case, $\frac{\partial l(\boldsymbol{\beta},\phi,\boldsymbol{\zeta}|\mathbf{y})}{\partial \boldsymbol{\eta}} \frac{\partial l(\boldsymbol{\beta},\phi,\boldsymbol{\zeta}|\mathbf{y})}{\partial \boldsymbol{\eta}'} = (\mathbf{y} - \boldsymbol{\eta})(\mathbf{y} - \boldsymbol{\eta})'/\sigma^2$ and $\frac{\partial^2 l(\boldsymbol{\beta},\phi,\boldsymbol{\zeta}|\mathbf{y})}{\partial \boldsymbol{\eta}\partial \boldsymbol{\eta}'} = -\mathbf{1}_N\mathbf{1}'_N/\sigma^2$. Therefore we have

$$B^{(1)}_{i,k} = \left\{(\mathbf{y}_i - \mathbf{X}_i\boldsymbol{\beta})' Z_{ik}\right\}^2 - Z'_{ik}Z_{ik}$$
$$B^{(2)}_{i,m,k} = (\mathbf{y}_i - \mathbf{X}_i\boldsymbol{\beta})' Z_{im}Z'_{ik}(\mathbf{y}_i - \mathbf{X}_i\boldsymbol{\beta}) - Z'_{ik}Z_{im},$$

where Z_{ik} denotes the kth column of $\mathbf{Z}_i$, and

$$L_0 = \exp\left\{-\frac{1}{2\sigma^2}\sum_{i=1}^{n}\sum_{j=1}^{n_i}(y_{ij} - \mathbf{x}'_{ij}\boldsymbol{\beta})^2\right\}.$$

When y_{ij} are 0–1 random variables, the logistic regression model can be obtained by the canonical link function $g(\pi_{ij}) = \log\frac{\pi_{ij}}{1-\pi_{ij}} = \eta_{ij} = \mathbf{x}'_{ij}\boldsymbol{\beta} + \mathbf{z}'_{ij}\boldsymbol{\zeta}_i, \phi = 1$, $b(\theta_{ij}) = \log(1 + \mathrm{e}^{\eta_{ij}}) = -\log(1 - \pi_{ij})$, hence $\frac{\partial l(\boldsymbol{\beta},\phi,\boldsymbol{\zeta}|\mathbf{y})}{\partial \boldsymbol{\eta}} \frac{\partial l(\boldsymbol{\beta},\phi,\boldsymbol{\zeta}|\mathbf{y})}{\partial \boldsymbol{\eta}'} = (\mathbf{y} - \boldsymbol{\pi})(\mathbf{y} - \boldsymbol{\pi})'$ and $\frac{\partial^2 l(\boldsymbol{\beta},\phi,\boldsymbol{\zeta}|\mathbf{y})}{\partial \boldsymbol{\eta}\partial \boldsymbol{\eta}'} = -\boldsymbol{\pi}(\mathbf{1}_N - \boldsymbol{\pi})'$. Then

$$B^{(1)}_{i,k} = \left\{(\mathbf{y}_i - \boldsymbol{\pi}_i)' Z_{ik}\right\}^2 - \boldsymbol{\pi}'_i\mathrm{DG}(Z_{ik}Z'_{ik})(\mathbf{1}_{n_i} - \boldsymbol{\pi}_i)$$
$$B^{(2)}_{i,m,k} = (\mathbf{y}_i - \boldsymbol{\pi}_i)' Z_{im}Z'_{ik}(\mathbf{y}_i - \boldsymbol{\pi}_i) - \boldsymbol{\pi}'_i\mathrm{DG}(Z_{im}Z'_{ik})(\mathbf{1}_{n_i} - \boldsymbol{\pi}_i),$$

where $\boldsymbol{\pi}_i = (\pi_{i1}, \ldots, \pi_{in_i})'$ with $\pi_{ij} = \exp(\mathbf{x}'_{ij}\boldsymbol{\beta})/\left(1 + \exp(\mathbf{x}'_{ij}\boldsymbol{\beta})\right)$, and

$$L_0 = \exp\left\{y_{ij}\log\frac{\pi_{ij}}{1 - \pi_{ij}} + \log(1 - \pi_{ij})\right\}.$$

Similarly, when y_{ij} are counts with mean λ_{ij}, the Poisson regression model can be obtained by the canonical link function $g(\lambda_{ij}) = \log\lambda_{ij} = \eta_{ij} = \mathbf{x}'_{ij}\boldsymbol{\beta} + \mathbf{z}'_{ij}\boldsymbol{\zeta}_i$, $\phi = 1$, $b(\theta_{ij}) = \mathrm{e}^{\eta_{ij}} = \lambda_{ij}$, $\frac{\partial l(\boldsymbol{\beta},\phi,\boldsymbol{\zeta}|\mathbf{y})}{\partial \boldsymbol{\eta}} \frac{\partial l(\boldsymbol{\beta},\phi,\boldsymbol{\zeta}|\mathbf{y})}{\partial \boldsymbol{\eta}'} = (\mathbf{y} - \boldsymbol{\lambda})(\mathbf{y} - \boldsymbol{\lambda})'$ and $\frac{\partial^2 l(\boldsymbol{\beta},\phi,\boldsymbol{\zeta}|\mathbf{y})}{\partial \boldsymbol{\eta}\partial \boldsymbol{\eta}'} = -\boldsymbol{\lambda}\mathbf{1}'_N$. Then we obtain that

$$B^{(1)}_{i,k} = \left\{(\mathbf{y}_i - \boldsymbol{\lambda}_i)' Z_{ik}\right\}^2 - \boldsymbol{\lambda}'_i\mathrm{DG}(Z_{ik}Z'_{ik})\mathbf{1}_{n_i}$$
$$B^{(2)}_{i,m,k} = (\mathbf{y}_i - \boldsymbol{\lambda}_i)' Z_{im}Z'_{ik}(\mathbf{y}_i - \boldsymbol{\lambda}_i) - \boldsymbol{\lambda}'_i\mathrm{DG}(Z_{im}Z'_{ik})\mathbf{1}_{n_i},$$

where $\boldsymbol{\lambda}_i = (\lambda_{i1}, \ldots, \lambda_{in_i})'$ with $\lambda_{ij} = \exp(\mathbf{x}'_{ij}\boldsymbol{\beta})$, and $L_0 = \exp\left(y_{ij}\log\lambda_{ij} - \lambda_{ij} - \log y_{ij}!\right)$.

Part II
Factor Analysis and Structural Equations Models

A Unified Approach to Two-Level Structural Equation Models and Linear Mixed Effects Models

Peter M. Bentler and Jiajuan Liang

1 Introduction

Two-level structural equation models (two-level SEM for simplicity) are widely used to analyze correlated clustered data (or two-level data) such as data collected from students (level-1 units) nested in different schools (level-2 units), or data collected from siblings (level-1 units) nested in different families (level-2 units). These data are usually collected by two sampling steps: randomly choosing some level-2 units; and then, randomly choosing some level-1 units from each chosen level-2 unit. Data collected in this way can be considered to be affected by two different random sources or random effects, namely, level-1 effects and level-2 effects. The substantive goal with such two-level data is to obtain theoretically meaningful and statistically adequate submodels for both the level-1 and level-2 effects. Realization of this main task consists of three steps: (1) set up an initial model with both level-1 and level-2 effects; (2) estimate the unknown model parameters; and (3) test the goodness-of-fit of the given model.

In the context of latent variable structural equation modeling, Step 1 is based on an initial understanding and substantive knowledge related to possible constructs or factors that may affect the observed data. Depending on the field, this may involve setting up a measurement model with observed indicators of one or more latent variables, as in a confirmatory factor analysis model, or possibly also a multivariate relations model in which some latent variables affect others, as in a typical structural equations model. Such models may be specified at each of the two

P.M. Bentler
University of California, Los Angeles Departments of Psychology and Statistics, Box 951563, Los Angeles, CA 90095-1563
bentler@ucla.edu

J. Liang
University of New Haven, College of Business, 300 Boston Post Road, West Haven, CT 06516
jliang@newhaven.edu

D. B. Dunson (ed.) *Random Effect and Latent Variable Model Selection*,
DOI: 10.1007/978-0-387-76721-5,

levels, where the model at both levels can be highly similar or completely different. Clearly the more substantive knowledge one can have, the better the chance to identify a suitable model for characterizing the data. Step 2 requires some estimation machinery, and in two-level SEM, this is typically based on the asymptotic statistical theory of maximum likelihood (ML) or generalized least squares (GLS), see, for example, Bentler et al. (2005). Step 3, testing goodness-of-fit in two-level SEM, uses the standard testing machinery associated with ML and GLS methodology, and provides evidence on how well the model proposed in step 1 can represent the observed data.

Approaches to the estimation of parameters and model evaluation in two-level SEM are similar to their counterparts in conventional (one-level) SEM. These methods are well known due to some popular statistical programs such as EQS (Bentler, 2006), LISREL (du Toit and du Toit, 2001) and M*plus* (Muthen and Muthen, 2004). The basic approach to model evaluation involves the chi-square statistic which compares a restricted SEM to a more general unrestricted or saturated model. Because a proposed model (the null hypothesis) may fail to be acceptable in large samples, alternative fit indices and other statistics have been proposed for model evaluation, see, Yuan and Bentler (2007a,b) for conventional SEM, and Bentler et al. (2005) and Yuan and Bentler (2003) for two-level SEM. As in conventional SEM, acceptance of the null hypothesis (i.e., the proposed model) in two-level SEM by a test statistic does not necessarily imply that the proposed model is the correct model. When a model fails to fit, it may be desirable to improve it using some model modification or selection criteria (see, e.g., Bozdogan, 1987; Cudeck and Browne, 1983; Sörbom, 1989).

An important characteristic of two-level data is that the two different types of factors (i.e., level-1 and level-2 factors) are assumed to be the only factors that affect the data. This characteristic is similar to that of data affected by mixed effects and random effects. By taking the fixed effects as the effects from level-2 factors and the random effects as the effects from level-1 factors, we will describe a unified approach to two-level linear SEM and linear mixed effects models (LMEM for simplicity) using the same model formulation. Some equivalence between multilevel SEM and LMEM has been already studied by a number of researchers such as Rovine and Molenaar (2000), Bauer (2003), Curran (2003), Skrondal and Rabe-Hesketh (2004), and Mehta and Neale (2005) from different points of view. In this chapter, we focus on estimating model parameters in view of the same model formulation.

This chapter is organized as follows. We introduce the general model formulation for two-level SEM and its relation to LMEM in Sect. 2. An EM algorithm for estimating model parameters for both two-level SEM and LMEM, and some asymptotic properties of the parameter estimator are given in Sect. 3. Applications of the EM algorithm are illustrated by examples in Sect. 4. Some discussion and comments are given in the last section.

2 Model Formulation

In modeling two-level data, it is usually assumed that both level-1 and level-2 observations, respectively, have the same dimension (e.g., Lee and Poon, 1998; Bentler and Liang, 2003; Liang and Bentler, 2004). This assumption is violated when additional measurements are taken from factors or latent variables at level-1. A typical example of this situation is that students (level-1 units) nested in different schools (level-2 units) may be given different number of scholastic tests to measure their ability (e.g., the math ability). Then the data collected from students' scores in the tests will have different dimensions across the schools. We will call this situation one of dimensional heterogeneity. Therefore, it is meaningful to allow both level-1 and level-2 observations to have different dimensions.

Suppose that the observed data are collected from a hierarchical sampling scheme: (1) randomly choose some level-2 units (such as different schools or families); and (2) randomly choose some level-1 units (such as students or family members) from each chosen level-2 unit. Let $\{\boldsymbol{y}_{gi} : p_g \times 1,\ i = 1, \ldots, N_g\}$ denote the observations from level-2 unit g. For example, $\boldsymbol{y}_{gi}$ may denote the observation from the ith student (level-1 unit) nested in the gth school (level-2 unit), and there are p_g tests given to students in the gth school, say, $g = 1, \ldots, G$. G is called the level-2 sample size and $\{N_g : g = 1, \ldots, G\}$ are called the level-1 sample sizes. N_g may be different for different g (i.e., an unbalanced sample design). Let $\{z_g : q_g \times 1,\ g = 1, \ldots, G\}$ denote the pure level-2 observations that are only observed from level-2 units. For example, z_g may denote the financial resources for the gth school, and there are q_g financial resources for the gth school. Then $\{\boldsymbol{y}_{gi} : p_g \times 1,\ i = 1, \ldots, N_g;\ z_g : q_g \times 1,\ g = 1, \ldots, G\}$ constitutes a set of observations from all responses in the population. The level-1 observations $\{\boldsymbol{y}_{gi} : p_g \times 1,\ i = 1, \ldots, N_g\}$ (for each fixed g) are usually not independent because different level-1 units nested in the same level-2 unit are affected by some common level-2 factors. The pure level-2 observation z_g is assumed to have the same effect on all level-1 units nested in the same level-2 unit g. For example, the financial resources for the gth school can be assumed to have the same effect on all students (level-1 units) nested in the same school (level-2 unit) g. Based on this viewpoint, we propose the following general formulation for two-level SEM:

$$\begin{pmatrix} z_g \\ \boldsymbol{y}_{gi} \end{pmatrix} = \begin{pmatrix} z_g \\ \boldsymbol{v}_g \end{pmatrix} + \begin{pmatrix} \mathbf{0} \\ \boldsymbol{v}_{gi} \end{pmatrix}, \tag{1}$$

for ML analysis with the assumptions:

(A1) The level-1 random vectors $\{\boldsymbol{v}_{gi} : p_g \times 1,\ i = 1, \ldots, N_g\}$ are independent for each fixed g and $\boldsymbol{v}_{gi} \sim N_{p_g}(\mathbf{0}, \boldsymbol{\Sigma}_{gW})$ for $g = 1, \ldots, G$, $\boldsymbol{\Sigma}_{gW} > \mathbf{0}$ (positive definite).

(A2) The level-2 random vectors $\{\boldsymbol{v}_g : p_g \times 1,\ g = 1, \ldots, G\}$ are independent and $\boldsymbol{v}_g \sim N_{p_g}(\boldsymbol{\mu}_{2g}, \boldsymbol{\Sigma}_{gB})$ with $\boldsymbol{\Sigma}_{gB} > \mathbf{0}$.

(A3) $\{z_g : q_g \times 1,\ g = 1, \ldots, G\}$ are independent level-2 observations and $z_g \sim N_{q_g}(\boldsymbol{\mu}_{1g}, \boldsymbol{\Sigma}_{gzz})$ with $\boldsymbol{\Sigma}_{gzz} > \mathbf{0}$.

(A4) The random vector $(z'_g, \tilde{v}'_g)'$ $\left((p_g + q_g) \times 1\right)$ has a joint multivariate normal distribution $N_{p_g+q_g}(\mu_g, \tilde{\Sigma}_{gB})$ with $\tilde{\Sigma}_{gB} > \mathbf{0}$ and

$$\mu_g = \begin{pmatrix} \mu_{1g} \\ \mu_{2g} \end{pmatrix}, \quad \tilde{\Sigma}_{gB} = \mathrm{cov}\begin{pmatrix} z_g \\ v_g \end{pmatrix} = \begin{pmatrix} \Sigma_{gzz} & \Sigma_{gzy} \\ \Sigma_{gyz} & \Sigma_{gB} \end{pmatrix}, \tag{2}$$

where $\Sigma_{gzy} = \Sigma'_{gyz} = \mathrm{cov}(z_g, v_g)$.

(A5) $\{z_g, v_g\}$ is uncorrelated with $\{v_{gi} : i = 1, \ldots, N_g\}$ for each fixed g.

In formulation (1), within-group (level-1) differences are reflected by a model for v_{gi} for each given g $(i = 1, \ldots, N_g)$ and we call such a model, a level-1 model, which may contain level-1 latent factors; between-group (level-2) differences are reflected by a model for v_g and a model for z_g $(g = 1, \ldots, G)$ and we call such models, level-2 models, which may contain level-2 latent factors. All latent factors are assumed to have normal distributions in the following context to derive the unified algorithm for estimating model parameters.

A nontrivial model under formulation (1) is one that restricts the means and covariances in assumptions (A1)–(A4). This implies that the means and covariance matrices in assumptions (A1)–(A4) depend on a common parameter vector θ (say, $r \times 1$). θ contains all model parameters from formulation (1) and we can write

$$\mu_g = \mu_g(\theta), \quad \Sigma_{gW} = \Sigma_{gW}(\theta), \quad \tilde{\Sigma}_{gB} = \tilde{\Sigma}_{gB}\ (\theta). \tag{3}$$

These matrices may be structured in particular ways as motivated by specific structural models, see Sect. 4. Liang and Bentler (2004) proposed an ML analysis for the model formulation (1) for the case of $p_g \equiv p$ and $q_g \equiv q$ and pointed out that formulation (1) includes the formulations for two-level SEM in McDonald and Goldstein (1989), Muthén (1989), Raudenbush (1995), Lee (1990), and Lee and Poon (1998). An analysis of the model defined by (1) consists of two major tasks:

(a) Estimate the model parameter vector θ from the available observations $\{y_{gi}, z_g\}$.
(b) Evaluate the goodness-of-fit after the parameter θ is estimated.

After a model expressed by (1) is set up, it is assumed that the model is identified. That is, the mean and covariance structures in (3) are uniquely determined by θ and vice versa. This implies that if there are two parameters θ_1 $(r \times 1)$ and θ_2 $(r \times 1)$ such that

$$\mu_g(\theta_1) = \mu_g(\theta_2), \quad \Sigma_{gW}(\theta_1) = \Sigma_{gW}(\theta_2), \quad \tilde{\Sigma}_{gB}\ (\theta_1) = \tilde{\Sigma}_{gB}\ (\theta_2)$$

for $g = 1, \ldots, G$, then $\theta_1 = \theta_2$. The complexity of a model defined by (1) may come from complicated level-1 models for the within variable v_{gi}, or from complicated level-2 models for the between variable v_g and the observable variable z_g, or from both complicated within and between models. When v_{gi}, v_g, and z_g are determined by measurement models such as factor analysis models (1) reduces to a usual SEM with dimensional heterogeneity.

The mean and covariance structures in (3) act as the null hypothesis for the model formulation (1). The common parameter vector $\boldsymbol{\theta}$ in (3) is assumed to contain r distinct individual model parameters. The saturated (or trivial) model associated with (3) is the case that $\boldsymbol{\theta}$ contains the maximum number of distinct individual model parameters R that is given by

$$R = \sum_{g=1}^{G} \left[p_g + q_g + p_g(p_g + 1)/2 + (p_g + q_g)(p_g + q_g + 1)/2 \right]. \tag{4}$$

A model-fit statistic for model formulation (1) is a test of the restricted model (3) with $r < R$ versus the saturated model with $r = R$.

A noteworthy case of the model in (1) is that it covers the linear mixed effects models (LMEM). An LMEM can be formulated (Meng, 1998) as

$$\boldsymbol{y}_g = \boldsymbol{X}'_g \boldsymbol{\beta} + \boldsymbol{Z}'_g \boldsymbol{b}_g + \boldsymbol{e}_g, \tag{5}$$

where $\boldsymbol{b}_g \sim N_q(\boldsymbol{0}, \boldsymbol{T})$ $(\boldsymbol{T} > \boldsymbol{0})$ and $\boldsymbol{e}_g \sim N_{n_g}(\boldsymbol{0}, \sigma^2 \boldsymbol{R}_g)$ $(\boldsymbol{R}_g > \boldsymbol{0})$ are uncorrelated. $\boldsymbol{y}_g$ $(n_g \times 1)$ is the response vector from the gth group, $\boldsymbol{X}_g$ $(p \times n_g)$ is the design matrix from fixed effects, $\boldsymbol{Z}_g$ $(q \times n_g)$ contains observations from random effects, $\boldsymbol{b}_g$ $(q \times 1)$ contains the random effects, and $\boldsymbol{e}_g$ $(n_g \times 1)$ contains the random errors. In the formulation given by (1), let $N_1 = N_2 = \ldots = N_G \equiv 1$, $\boldsymbol{z}_g \equiv \boldsymbol{0}$ (no level-2 observations), and

$$\boldsymbol{v}_g = \boldsymbol{X}'_g \boldsymbol{\beta} + \boldsymbol{Z}'_g \boldsymbol{b}_g, \quad \boldsymbol{v}_{gi} \equiv \boldsymbol{e}_g \quad (i \equiv 1) \tag{6}$$

then $\boldsymbol{y}_{gi} \equiv \boldsymbol{y}_g$ $(i \equiv 1)$ and

$$\boldsymbol{y}_{gi} = \boldsymbol{v}_g + \boldsymbol{v}_{gi} \qquad (i \equiv 1). \tag{7}$$

It is noted that to fit the LMEM (5) within the SEM framework (1), we have to switch the meaning of dimensions: in formulation (1), for each given g, N_g stands for the number of individuals within group g; but when taking $N_1 = N_2 = \ldots = N_G \equiv 1$ in fitting the LMEM (5) into (7), each N_g only stands for an imaginary level-1 sample size. Taking $N_1 = N_2 = \ldots = N_G \equiv 1$ in (5) does not mean that the number of individuals within groups are all equal to 1. The number of individuals within groups in LMEM (5) is actually n_g $(g = 1, \ldots, G)$, which is now switched to the dimension or the number of measured outcomes that is equivalent to p_g in formulation (1), which is the dimension of level-1 observations in a two-level SEM expressed by (1). A similar switch of the meaning of dimension can be also noticed in papers that provide a unifying framework for LMEM and multilevel SEM, see, for example, Muthén (1997), Bauer (2003), Curran (2003), Skrondal and Rabe-Hesketh (2004). Since an LMEM is essentially a single-level model, there are no second-level observations except the individual observations. The term $\boldsymbol{z}_g$ representing a level-2 observation in formulation (1) becomes imaginary in an LMEM. Taking $\boldsymbol{z}_g \equiv \boldsymbol{0}$ in fitting the LMEM (5) into formulation (1) simply means that there are no level-2 observations in LMEM and so formulation (1) reduces to (7), which is a special case of formulation (1) without level-2 observations.

If we assume $n_g \leq q$ for the full-rank design matrix $\boldsymbol{Z}_g$ in (5), then $\boldsymbol{v}_g = \boldsymbol{X}'_g\boldsymbol{\beta} + \boldsymbol{Z}'_g\boldsymbol{b}_g$ has a nonsingular multinormal distribution and the assumptions (A1)–(A5) will be satisfied for the model defined by (7). In the LMEM (5), $\boldsymbol{b}_g \sim N_q(\boldsymbol{0}, \boldsymbol{T})$ can be considered as a between (level-2) latent variable and $\boldsymbol{e}_g \sim N_{n_g}(\boldsymbol{0}, \sigma^2\boldsymbol{R}_g)$ as a within (level-1) latent variable. $\boldsymbol{y}_{gi} \equiv \boldsymbol{y}_g$ $(i \equiv 1)$ is the only one response (observation) vector from group (level-2 unit) g. Therefore the formulation defined by (1) covers all LMEM of the type defined by (5) with all $n_g \leq q$.

Let $\boldsymbol{\theta}$ be the model parameter vector in (5) that is composed of all free model parameters in (5): the regression coefficients in $\boldsymbol{\beta}$, the variances and nonduplicated covariances in $\text{cov}(\boldsymbol{b}_g) = \boldsymbol{T}$, and the variances and nonduplicated covariances in $\text{cov}(\boldsymbol{e}_g) = \sigma^2\boldsymbol{R}_g$. The mean and covariance structures (the restricted model or null hypothesis) associated with the LMEM (5) (or (7)) are given by

$$\boldsymbol{\mu}_g(\boldsymbol{\theta}) = E(\boldsymbol{y}_g) = \boldsymbol{X}'_g\boldsymbol{\beta} + \boldsymbol{Z}'_g E(\boldsymbol{b}_g) = \boldsymbol{X}'_g\boldsymbol{\beta}, \quad \boldsymbol{\Sigma}_{gW}(\boldsymbol{\theta}) = \text{cov}(\boldsymbol{e}_g) = \sigma^2\boldsymbol{R}_g(\boldsymbol{\theta}),$$
$$\boldsymbol{\Sigma}_{gB}(\boldsymbol{\theta}) = \text{cov}(\boldsymbol{X}'_g\boldsymbol{\beta} + \boldsymbol{Z}'_g\boldsymbol{b}_g) = \boldsymbol{Z}'_g\text{cov}(\boldsymbol{b}_g)\boldsymbol{Z}_g = \boldsymbol{Z}'_g\boldsymbol{T}(\boldsymbol{\theta})\boldsymbol{Z}_g. \tag{8}$$

The saturated model (alternative hypothesis) associated with the LMEM (5) is usually far more complicated than that associated with the two-level SEM expressed by formulation (1) because there may be far too many different within-group covariance matrices. It has been noted that there is no estimable saturated model for many LMEM's (especially for highly imbalanced data). In the following context, we will only focus on a unified approach to estimating model parameters for both two-level SEM and LMEM under the structured models (the null hypothesis is true or the presented structural relationships are assumed to be correct).

3 The EM Algorithm

3.1 Maximum Likelihood Estimation

Because the general linear mixed effects model defined by (5) is a special case of the model formulation given by (1), in this section we will develop an EM algorithm for computing the MLE (maximum likelihood estimate) of the model parameter vector $\boldsymbol{\theta}$ specified in the mean and covariance structures in (3). Then the EM algorithm can be applied to the LMEM (5). Let $\boldsymbol{\theta}$ $(r \times 1)$ be the parameter vector containing all model parameters from formulation (1) associated with assumptions (A1)–(A5).

To simplify the derivation of the EM algorithm, let

$$\boldsymbol{y}_g = \begin{pmatrix} \boldsymbol{y}_{g1} \\ \vdots \\ \boldsymbol{y}_{gN_g} \end{pmatrix}, \quad \boldsymbol{Y}_{g0} = \begin{pmatrix} \boldsymbol{y}'_{g1} \\ \vdots \\ \boldsymbol{y}'_{gN_g} \end{pmatrix}, \quad \boldsymbol{u}_g = \begin{pmatrix} \boldsymbol{z}_g \\ \bar{\boldsymbol{y}}_g \end{pmatrix}, \quad \boldsymbol{x}_g = \begin{pmatrix} \boldsymbol{v}_g \\ \boldsymbol{u}_g \end{pmatrix},$$
$$\boldsymbol{X} = \{\boldsymbol{x}_1, \ldots, \boldsymbol{x}_G\}, \quad \boldsymbol{Z} = \{\boldsymbol{z}_1, \ldots, \boldsymbol{z}_G\}, \quad \bar{\boldsymbol{y}}_g = \frac{1}{N_g}\sum_{i=1}^{N_g} \boldsymbol{y}_{gi}. \tag{9}$$

Then $\boldsymbol{u}_g$ $\left[(q_g + N_g p_g) \times 1\right]$ contains all observations from group (level-2 unit) g. Taking $\boldsymbol{v}_g$ as a missing vector value, we construct the "complete observation" vector $\boldsymbol{x}_g$ $\left[(q_g + p_g + N_g p_g) \times 1\right]$. $\boldsymbol{Y}_{g0}$ in (9) is the observation matrix composed of all level-1 observations from group g. $\bar{\boldsymbol{y}}_g$ is the sample mean from the level-1 observations in group g. $\boldsymbol{X}$ denotes the set of all complete observations and $\boldsymbol{Z}$ the set of all level-2 observations. From the assumptions (A1)–(A5) on (1), we can derive the negative twice the logarithm of the likelihood function from all complete responses $\{\boldsymbol{x}_g : g = 1, \ldots, G\}$

$$l(\boldsymbol{X}, \boldsymbol{\theta}^*) = \sum_{g=1}^{G} l_g(\boldsymbol{X}, \boldsymbol{\theta}^*), \tag{10}$$

where $\boldsymbol{\theta}^*$ is an arbitrarily specified value of the parameter $\boldsymbol{\theta}$, and

$$\begin{aligned} l_g(\boldsymbol{X}, \boldsymbol{\theta}^*) &= \log|\boldsymbol{\Sigma}^*_{gB}| + (\boldsymbol{v}_g - \boldsymbol{\mu}^*_{2g})'\boldsymbol{\Sigma}^{*-1}_{gB}(\boldsymbol{v}_g - \boldsymbol{\mu}^*_{2g}) \\ &+ \sum_{i=1}^{N_g}\left\{\log|\boldsymbol{\Sigma}^*_{gW}| + (\boldsymbol{y}_{gi} - \boldsymbol{v}_g)'\boldsymbol{\Sigma}^{*-1}_{gW}(\boldsymbol{y}_{gi} - \boldsymbol{v}_g)\right\} \\ &+ \log|\boldsymbol{\Sigma}^*_{gzz.v}| + \left[\boldsymbol{z}_g - \boldsymbol{\mu}^*_{1g} - \boldsymbol{\Sigma}^*_{gzy}\boldsymbol{\Sigma}^{*-1}_{gB}(\boldsymbol{v}_g - \boldsymbol{\mu}^*_{2g})\right]' \\ &\times \boldsymbol{\Sigma}^{*-1}_{gzz.v}\left[\boldsymbol{z}_g - \boldsymbol{\mu}^*_{1g} - \boldsymbol{\Sigma}^*_{gzy}\boldsymbol{\Sigma}^{*-1}_{gB}(\boldsymbol{v}_g - \boldsymbol{\mu}^*_{2g})\right], \end{aligned} \tag{11}$$

where

$$\begin{aligned} &\boldsymbol{\mu}^*_{1g} = \boldsymbol{\mu}_{1g}(\boldsymbol{\theta}^*), \quad \boldsymbol{\mu}^*_{2g} = \boldsymbol{\mu}_{2g}(\boldsymbol{\theta}^*), \quad \boldsymbol{\Sigma}^*_{gW} = \boldsymbol{\Sigma}_{gW}(\boldsymbol{\theta}^*), \\ &\boldsymbol{\Sigma}^*_{gB} = \boldsymbol{\Sigma}_{gB}(\boldsymbol{\theta}^*), \quad \boldsymbol{\Sigma}^*_{gzz} = \boldsymbol{\Sigma}_{gzz}(\boldsymbol{\theta}^*), \quad \boldsymbol{\Sigma}^*_{gzy} = \boldsymbol{\Sigma}_{gzy}(\boldsymbol{\theta}^*), \\ &\boldsymbol{\Sigma}^*_{gyz} = \boldsymbol{\Sigma}^{*\prime}_{gzy}, \quad \boldsymbol{\Sigma}^*_{gzz.v} = \boldsymbol{\Sigma}^*_{gzz} - \boldsymbol{\Sigma}^*_{gzy}\boldsymbol{\Sigma}^{*-1}_{gB}\boldsymbol{\Sigma}^*_{gyz}. \end{aligned} \tag{12}$$

According to the principle of the EM algorithm (Dempster et al., 1977), the E-step function of the EM algorithm for estimating $\boldsymbol{\theta}$ is the conditional expectation defined by

$$M(\boldsymbol{\theta}^*|\boldsymbol{\theta}) = \sum_{g=1}^{G} E\left\{l_g(\boldsymbol{X}, \boldsymbol{\theta}^*)|\boldsymbol{z}_g, \boldsymbol{y}_g, \boldsymbol{\theta}\right\}, \tag{13}$$

where both $\boldsymbol{\theta}^*$ and $\boldsymbol{\theta}$ are two arbitrarily specified values of the same parameter $\boldsymbol{\theta}$. From assumptions (A1)–(A5) on (1), it can be proved that

$$\begin{aligned} E\left\{l_g(\boldsymbol{X}, \boldsymbol{\theta}^*)|\boldsymbol{z}_g, \boldsymbol{y}_g, \boldsymbol{\theta}\right\} = &\left\{\log|\boldsymbol{\Sigma}^*_{gB}| + \mathrm{tr}\left(\boldsymbol{\Sigma}^{*-1}_{gB}\boldsymbol{S}^*_{gB}\right)\right\} \\ &+ N_g\left\{\log|\boldsymbol{\Sigma}^*_{gW}| + \mathrm{tr}\left(\boldsymbol{\Sigma}^{*-1}_{gW}\boldsymbol{S}_{gW}\right)\right\} \\ &+ \left\{\log|\boldsymbol{\Sigma}^*_{gzz.v}| + \mathrm{tr}\left(\boldsymbol{\Sigma}^{*-1}_{gzz.v}\boldsymbol{S}^*_{gz}\right)\right\}, \end{aligned} \tag{14}$$

where

$$
\begin{aligned}
S^*_{gB} &= E\Big[(\boldsymbol{v}_g - \boldsymbol{\mu}^*_{2g})(\boldsymbol{v}_g - \boldsymbol{\mu}^*_{2g})'|\boldsymbol{z}_g, \boldsymbol{y}_g, \boldsymbol{\theta}\Big],\\
S_{gW} &= \frac{1}{N_g}\sum_{i=1}^{N_g} E\Big[(\boldsymbol{y}_{gi} - \boldsymbol{v}_g)(\boldsymbol{y}_{gi} - \boldsymbol{v}_g)'|\boldsymbol{z}_g, \boldsymbol{y}_g, \boldsymbol{\theta}\Big],\\
S^*_{gz} &= E\Big\{\Big[\boldsymbol{z}_g - \boldsymbol{\mu}^*_{1g} - \boldsymbol{\Sigma}^*_{gzy}\boldsymbol{\Sigma}^{*-1}_{gB}(\boldsymbol{v}_g - \boldsymbol{\mu}^*_{2g})\Big]\\
&\quad \times\Big[\boldsymbol{z}_g - \boldsymbol{\mu}^*_{1g} - \boldsymbol{\Sigma}^*_{gzy}\boldsymbol{\Sigma}^{*-1}_{gB}(\boldsymbol{v}_g - \boldsymbol{\mu}^*_{2g})\Big]'|\boldsymbol{z}_g, \boldsymbol{y}_g, \boldsymbol{\theta}\Big\}.
\end{aligned} \tag{15}
$$

Under the assumptions (A1)–(A5) on (1), it can be derived that

$$
\begin{aligned}
\boldsymbol{a}_g(\boldsymbol{\theta}) = \boldsymbol{a}_g &\overset{\text{def}}{=} E(\boldsymbol{v}_g|\boldsymbol{z}_g, \boldsymbol{y}_g, \boldsymbol{\theta}) = \boldsymbol{\mu}_{2g} + \boldsymbol{\Sigma}_{1g}\boldsymbol{\Omega}_g^{-1}(\boldsymbol{c}_g - \boldsymbol{\mu}_g),\\
\boldsymbol{C}_g(\boldsymbol{\theta}) = \boldsymbol{C}_g &\overset{\text{def}}{=} \operatorname{cov}(\boldsymbol{v}_g|\boldsymbol{z}_g, \boldsymbol{y}_g, \boldsymbol{\theta}) = \boldsymbol{\Sigma}_{gB} - \boldsymbol{\Sigma}_{1g}\boldsymbol{\Omega}_g^{-1}\boldsymbol{\Sigma}'_{1g},
\end{aligned} \tag{16}
$$

where "$\overset{\text{def}}{=}$" means "defined as," $\boldsymbol{\mu}_g$ is defined in (2), and

$$
\begin{aligned}
\boldsymbol{\Sigma}_{1g} &= (\boldsymbol{\Sigma}_{gyz}, \boldsymbol{\Sigma}_{gB}), \qquad \boldsymbol{\Sigma}_g = \boldsymbol{\Sigma}_{gW} + N_g\boldsymbol{\Sigma}_{gB},\\
\boldsymbol{c}_g &= \begin{pmatrix} \boldsymbol{z}_g \\ \bar{\boldsymbol{y}}_g \end{pmatrix}, \qquad \boldsymbol{\Omega}_g = \begin{pmatrix} \boldsymbol{\Sigma}_{gzz} & \boldsymbol{\Sigma}_{gzy} \\ \boldsymbol{\Sigma}_{gyz} & \frac{1}{N_g}\boldsymbol{\Sigma}_g \end{pmatrix}.
\end{aligned} \tag{17}
$$

Then we have

$$
\begin{aligned}
S^*_{gB} &= \boldsymbol{C}_g + \boldsymbol{a}_g\boldsymbol{a}'_g - \boldsymbol{a}_g\boldsymbol{\mu}^{*\prime}_{2g} - (\boldsymbol{a}_g\boldsymbol{\mu}^{*\prime}_{2g})' + \boldsymbol{\mu}^*_{2g}\boldsymbol{\mu}^{*\prime}_{2g},\\
S_{gW} &= S_{gyy} + \boldsymbol{C}_g + \boldsymbol{a}_g\boldsymbol{a}'_g - \boldsymbol{a}_g\bar{\boldsymbol{y}}'_g - (\boldsymbol{a}_g\bar{\boldsymbol{y}}'_g)',\\
S^*_{gz} &= (\boldsymbol{z}_g - \boldsymbol{\mu}^*_{1g})(\boldsymbol{z}_g - \boldsymbol{\mu}^*_{1g})' - \boldsymbol{\Sigma}^*_{gzy}\boldsymbol{\Sigma}^{*-1}_{gB}(\boldsymbol{a}_g - \boldsymbol{\mu}^*_{2g})(\boldsymbol{z}_g - \boldsymbol{\mu}^*_{1g})'\\
&\quad -\Big[\boldsymbol{\Sigma}^*_{gzy}\boldsymbol{\Sigma}^{*-1}_{gB}(\boldsymbol{a}_g - \boldsymbol{\mu}^*_{2g})(\boldsymbol{z}_g - \boldsymbol{\mu}^*_{1g})'\Big]' + \boldsymbol{\Sigma}^*_{gzy}\boldsymbol{\Sigma}^{*-1}_{gB}S^*_{gB}\boldsymbol{\Sigma}^{*-1}_{gB}\boldsymbol{\Sigma}^*_{gyz}
\end{aligned} \tag{18}
$$

and

$$
S_{gyy} = \frac{1}{N_g}\sum_{i=1}^{N_g}\boldsymbol{y}_{gi}\boldsymbol{y}'_{gi} = \frac{1}{N_g}\boldsymbol{Y}'_{g0}\boldsymbol{Y}_{g0}. \tag{19}
$$

Employing some properties for block matrices, we can derive

$$
\begin{aligned}
M(\boldsymbol{\theta}^*|\boldsymbol{\theta}) &= \sum_{g=1}^{G} N_g\left\{\log|\boldsymbol{\Sigma}^*_{gW}| + \operatorname{tr}\left(\boldsymbol{\Sigma}^{*-1}_{gW}S_{gW}\right)\right\}\\
&\quad + \sum_{g=1}^{G}\left\{\log|\tilde{\boldsymbol{\Sigma}}^*_g| + \operatorname{tr}\left(\tilde{\boldsymbol{\Sigma}}^{*-1}_g\tilde{S}_g\right)\right\},
\end{aligned} \tag{20}
$$

where S_{gW} is given in (16), and

$$\tilde{\boldsymbol{\Sigma}}_g^* = \begin{pmatrix} \tilde{\boldsymbol{\Sigma}}_{gB}^* + \boldsymbol{\mu}_g^* \boldsymbol{\mu}_g^{*\prime} & \boldsymbol{\mu}_g^* \\ \boldsymbol{\mu}_g^{*\prime} & 1 \end{pmatrix}, \quad \tilde{\boldsymbol{S}}_g = \begin{pmatrix} \tilde{\boldsymbol{S}}_{gB} + \boldsymbol{d}_g \boldsymbol{d}_g' & \boldsymbol{d}_g \\ \boldsymbol{d}_g' & 1 \end{pmatrix}, \quad \boldsymbol{d}_g = \begin{pmatrix} \boldsymbol{z}_g \\ \boldsymbol{a}_g \end{pmatrix},$$

$$\tilde{\boldsymbol{S}}_{gB} = \begin{pmatrix} \boldsymbol{0} & \boldsymbol{0} \\ \boldsymbol{0} & \boldsymbol{C}_g \end{pmatrix}, \quad \boldsymbol{\mu}_g^* = \boldsymbol{\mu}_g(\boldsymbol{\theta}^*), \quad \tilde{\boldsymbol{\Sigma}}_{gB}^* = \tilde{\boldsymbol{\Sigma}}_{gB}(\boldsymbol{\theta}^*) \tag{21}$$

with $\boldsymbol{\mu}_g$ and $\tilde{\boldsymbol{\Sigma}}_{gB}$ defined in (2) and $\boldsymbol{a}_g$ in (14). Here we express the matrices $\tilde{\boldsymbol{\Sigma}}_g^*$ and $\tilde{\boldsymbol{S}}_g$ as in (19) for programming convenience, since the expression (18) has exactly the same form as the multiple-group likelihood function in covariance structure analysis. The constant "1" in the matrices $\tilde{\boldsymbol{\Sigma}}_g^*$ and $\tilde{\boldsymbol{S}}_g$ in (19) is added to the E-step function (18) without loss of generality because $M_g(\boldsymbol{\theta}^*|\boldsymbol{\theta})$ and $M_g(\boldsymbol{\theta}^*|\boldsymbol{\theta}) \pm 1$ have the same optimality. The expressions for $\tilde{\boldsymbol{\Sigma}}_g^*$ and $\tilde{\boldsymbol{S}}_g$ in (19) with the constant "1" are the result of re-combination of smaller matrices into bigger matrices by using the property of block matrices

$$\tilde{\boldsymbol{\Sigma}}_g^{*-1} = \begin{pmatrix} \tilde{\boldsymbol{\Sigma}}_{gB}^* + \boldsymbol{\mu}_g^* \boldsymbol{\mu}_g^{*\prime} & \boldsymbol{\mu}_g^* \\ \boldsymbol{\mu}_g^{*\prime} & 1 \end{pmatrix}^{-1} = \begin{pmatrix} \tilde{\boldsymbol{\Sigma}}_{gB}^{*-1} & -\tilde{\boldsymbol{\Sigma}}_{gB}^{*-1} \boldsymbol{\mu}_g^* \\ -\boldsymbol{\mu}_g^{*\prime} \tilde{\boldsymbol{\Sigma}}_{gB}^{*-1} & 1 + \boldsymbol{\mu}_g^{*\prime} \tilde{\boldsymbol{\Sigma}}_{gB}^{*-1} \boldsymbol{\mu}_g^* \end{pmatrix},$$

$$\left|\tilde{\boldsymbol{\Sigma}}_g^*\right| = \begin{vmatrix} \tilde{\boldsymbol{\Sigma}}_{gB}^* + \boldsymbol{\mu}_g^* \boldsymbol{\mu}_g^{*\prime} & \boldsymbol{\mu}_g^* \\ \boldsymbol{\mu}_g^{*\prime} & 1 \end{vmatrix} = \left|\tilde{\boldsymbol{\Sigma}}_{gB}^*\right|,$$

$$\mathrm{tr}\left(\boldsymbol{\Sigma}_g^{*-1} \tilde{\boldsymbol{S}}_g\right) = \mathrm{tr}\left(\boldsymbol{\Sigma}_{gB}^{*-1} \tilde{\boldsymbol{S}}_{gB}\right) + \boldsymbol{d}_g' \tilde{\boldsymbol{\Sigma}}_{gB}^{*-1} \boldsymbol{d}_g - 2\boldsymbol{\mu}_g^{*\prime} \tilde{\boldsymbol{\Sigma}}_{gB}^{*-1} \boldsymbol{d}_g + \boldsymbol{\mu}_g^{*\prime} \tilde{\boldsymbol{\Sigma}}_{gB}^{*-1} \boldsymbol{\mu}_g^* + 1.$$

The M-step of the EM algorithm is to update $\boldsymbol{\theta}^*$ for every given $\boldsymbol{\theta}$ in the E-step function in (18). For example, in the ith step ($i = 0$ corresponds to the initial step), given $\boldsymbol{\theta} = \boldsymbol{\theta}_i$, we need to update $\boldsymbol{\theta}^*$ to $\boldsymbol{\theta}_{i+1}$ such that

$$M(\boldsymbol{\theta}_{i+1}|\boldsymbol{\theta}_i) \leq M(\boldsymbol{\theta}_i|\boldsymbol{\theta}_i) \tag{22}$$

according to the general idea of the EM-type algorithm (McLachlan and Krishnan, 1997). We will employ Lange's (1995a,b) EM gradient algorithm to derive the updating formula for $\boldsymbol{\theta}_{i+1}$ in (20) by using the gradient direction. This requires the first-order derivative of the E-step function (18) and a suitable approximation of the Fisher information matrix (McLachlan and Krishnan, 1997). The first-order derivative can be derived as

$$dM(\boldsymbol{\theta}|\boldsymbol{\theta}) = \left.\frac{\partial M(\boldsymbol{\theta}^*|\boldsymbol{\theta})}{\partial \boldsymbol{\theta}^*}\right|_{\boldsymbol{\theta}^*=\boldsymbol{\theta}}$$

$$= \sum_{g=1}^{G} \left\{ N_g \boldsymbol{\Delta}_{gW} \left(\boldsymbol{\Sigma}_{gW}^{-1} \otimes \boldsymbol{\Sigma}_{gW}^{-1} \right) \operatorname{vec}\left(\boldsymbol{\Sigma}_{gW} - \boldsymbol{S}_{gW} \right) \right\} \tag{23}$$

$$+ \textstyle\sum_{g=1}^{G} \left\{ \tilde{\boldsymbol{\Delta}}_g \left(\tilde{\boldsymbol{\Sigma}}_g^{-1} \otimes \tilde{\boldsymbol{\Sigma}}_g^{-1} \right) \operatorname{vec}\left(\tilde{\boldsymbol{\Sigma}}_g - \tilde{\boldsymbol{S}}_g \right) \right\},$$

where

$$\boldsymbol{\Delta}_{gW} = \frac{\partial(\operatorname{vec}\boldsymbol{\Sigma}_{gW})'}{\partial \boldsymbol{\theta}}, \quad \tilde{\boldsymbol{\Delta}}_g = \frac{\partial(\operatorname{vec}\tilde{\boldsymbol{\Sigma}}_g)'}{\partial \boldsymbol{\theta}}, \quad \tilde{\boldsymbol{\Sigma}}_g = \left.\tilde{\boldsymbol{\Sigma}}_g^*\right|_{\boldsymbol{\theta}^*=\boldsymbol{\theta}} \tag{24}$$

the sign "vec" in (21) and (22) stands for the vectorization of a matrix by stacking its columns successively, and the sign "⊗" for the Kronecker product of matrices. By using a positive definite matrix of first-order derivatives to approximate the Hessian matrix (see, e.g., (McLachlan and Krishnan, 1997), pp. 5–7, for the use of a positive definite matrix to approximate the Hessian matrix), we can obtain the following approximation to the Hessian matrix

$$I(\boldsymbol{\theta}) = E\left\{ \left.\frac{\partial^2 M(\boldsymbol{\theta}^*|\boldsymbol{\theta})}{\partial \boldsymbol{\theta}^* \partial \boldsymbol{\theta}^{*\prime}}\right|_{\boldsymbol{\theta}^*=\boldsymbol{\theta}} \right\}$$

$$\approx \sum_{g=1}^{G} N_g \boldsymbol{\Delta}_{gW} \left\{ 2\left(\boldsymbol{\Sigma}_{gW}^{-1} \boldsymbol{S}_{gew} \boldsymbol{\Sigma}_{gW}^{-1} \right) \otimes \boldsymbol{\Sigma}_{gW}^{-1} - \boldsymbol{\Sigma}_{gW}^{-1} \otimes \boldsymbol{\Sigma}_{gW}^{-1} \right\} \boldsymbol{\Delta}_{gW}' \tag{25}$$

$$+ \sum_{g=1}^{G} \tilde{\boldsymbol{\Delta}}_g \left\{ 2\left(\tilde{\boldsymbol{\Sigma}}_g^{-1} \tilde{\boldsymbol{S}}_{geb} \tilde{\boldsymbol{\Sigma}}_g^{-1} \right) \otimes \tilde{\boldsymbol{\Sigma}}_g^{-1} - \tilde{\boldsymbol{\Sigma}}_g^{-1} \otimes \tilde{\boldsymbol{\Sigma}}_g^{-1} \right\} \tilde{\boldsymbol{\Delta}}_g'.$$

The two matrices $\boldsymbol{S}_{gew}$ and $\tilde{\boldsymbol{S}}_{geb}$ are derived by the assumptions (A1)–(A5) on (1)

$$\begin{aligned} \boldsymbol{S}_{egw} &= E(\boldsymbol{S}_{gw}) = 2\boldsymbol{\Sigma}_{gB} + \boldsymbol{\Sigma}_{gW} - \boldsymbol{A}_{gw} - \boldsymbol{A}_{gw}', \\ \boldsymbol{A}_{gw} &= \boldsymbol{\Sigma}_{2g} \boldsymbol{\Omega}_g^{-1} \boldsymbol{\Sigma}_{1g}', \quad \boldsymbol{\Sigma}_{2g} = (\boldsymbol{\Sigma}_{gyz}, \tfrac{1}{N_g}\boldsymbol{\Sigma}_g), \end{aligned} \tag{26}$$

where $\boldsymbol{\Sigma}_{1g}$, $\boldsymbol{\Sigma}_g$ and $\boldsymbol{\Omega}_g$ are defined in (15), and

$$\begin{aligned} \tilde{\boldsymbol{S}}_{geb} &= E(\tilde{\boldsymbol{S}}_g) = \begin{pmatrix} \boldsymbol{S}_{geb} + \boldsymbol{\mu}_g \boldsymbol{\mu}_g' & \boldsymbol{\mu}_g \\ \boldsymbol{\mu}_g' & 1 \end{pmatrix}, \quad \boldsymbol{S}_{geb} = \begin{pmatrix} \boldsymbol{\Sigma}_{gzz} & \boldsymbol{A}_{gb} \\ \boldsymbol{A}_{gb}' & \boldsymbol{\Sigma}_{gB} \end{pmatrix}, \\ \boldsymbol{A}_{gb} &= (\boldsymbol{\Sigma}_{gzz}, \boldsymbol{\Sigma}_{gzy}) \boldsymbol{\Omega}_g^{-1} \boldsymbol{\Sigma}_{1g}'. \end{aligned} \tag{27}$$

According to the EM gradient algorithm in Lange (1995a,b), the M-step in the EM algorithm for estimating the parameter $\boldsymbol{\theta}$ can be realized by

$$\boldsymbol{\theta}_{i+1} = \boldsymbol{\theta}_i - \alpha I(\boldsymbol{\theta}_i)^{-1} dM(\boldsymbol{\theta}_i|\boldsymbol{\theta}_i), \tag{28}$$

where $\boldsymbol{\theta}_i$ denotes the value of $\boldsymbol{\theta}$ at the ith iteration and $0 < \alpha \leq 1$ is an adjusting constant for controlling the step length during the iteration. α can be chosen dynamically (it could be different) at each iteration. The Root Mean Square Error (RMSE) (Lee and Poon, 1998) can be used as a stopping criterion for the iteration in (26). That is, the iteration stops when the RMSE between two adjacent steps is small enough

$$\text{RMSE}\,(\boldsymbol{\theta}_{i+1}, \boldsymbol{\theta}_i) = \left\{ \frac{1}{r} \|\boldsymbol{\theta}_{i+1} - \boldsymbol{\theta}_i\|^2 \right\}^{1/2} \leq \epsilon \quad (\text{e.g., } \epsilon = 10^{-6}), \tag{29}$$

where the sign "$\|\cdot\|$" stands for the usual Euclidean distance, and r is the dimension of $\boldsymbol{\theta}$.

For the linear mixed effects model given by (5)–(7) with mean and covariance structures (8), we can obtain the first-order derivatives

$$\begin{aligned} \boldsymbol{\Delta}_{g\mu} &= \frac{\partial \boldsymbol{\mu}_g'}{\partial \boldsymbol{\theta}} = \frac{\partial \boldsymbol{\beta}'}{\partial \boldsymbol{\theta}} \cdot \boldsymbol{X}_g, \quad \boldsymbol{\Delta}_{gW} = \frac{\partial (\text{vec}\boldsymbol{\Sigma}_{gW})'}{\partial \boldsymbol{\theta}} = \frac{\partial \left[\text{vec}\left(\sigma^2 \boldsymbol{R}_g\right)\right]'}{\partial \boldsymbol{\theta}}, \\ \boldsymbol{\Delta}_{gB} &= \frac{\partial (\text{vec}\boldsymbol{\Sigma}_{gB})'}{\partial \boldsymbol{\theta}} = \frac{\partial (\text{vec}\boldsymbol{T})'}{\partial \boldsymbol{\theta}} \cdot (\boldsymbol{Z}_g \otimes \boldsymbol{Z}_g'). \end{aligned} \tag{30}$$

These derivatives are helpful in computing the term $\tilde{\boldsymbol{\Delta}}_g$ in (21)–(23). Because $N_g \equiv 1$ and $z_g \equiv \mathbf{0}$ in (1) $(g = 1, \ldots, G)$ for the LMEM (5)–(7), we can obtain the simplified formulas for $dM(\boldsymbol{\theta}|\boldsymbol{\theta})$ (see (21)) and $I(\boldsymbol{\theta})$ (see (23)) used in the iteration process (26)

$$\begin{aligned} dM(\boldsymbol{\theta}|\boldsymbol{\theta}) &= \frac{1}{\sigma^4} \sum_{g=1}^{G} \boldsymbol{\Delta}_{gW} \left(\boldsymbol{R}_g^{-1} \otimes \boldsymbol{R}_g^{-1}\right) \text{vec}\left(\sigma^2 \boldsymbol{R}_g - \boldsymbol{S}_{gW}\right) \\ &\quad + \sum_{g=1}^{G} \tilde{\boldsymbol{\Delta}}_{gm} \left(\tilde{\boldsymbol{\Sigma}}_{gm}^{-1} \otimes \tilde{\boldsymbol{\Sigma}}_{gm}^{-1}\right) \text{vec}\left(\tilde{\boldsymbol{\Sigma}}_{gm} - \tilde{\boldsymbol{S}}_{gm}\right), \\ I(\boldsymbol{\theta}) &= \sum_{g=1}^{G} \left\{ \frac{1}{\sigma^4} \boldsymbol{\Delta}_{gW} \left(\boldsymbol{R}_g^{-1} \otimes \boldsymbol{R}_g^{-1}\right) \boldsymbol{\Delta}_{gW}' + \tilde{\boldsymbol{\Delta}}_{gm} \left(\tilde{\boldsymbol{\Sigma}}_{gm}^{-1} \otimes \tilde{\boldsymbol{\Sigma}}_{gm}^{-1}\right) \tilde{\boldsymbol{\Delta}}_{gm}' \right\}. \end{aligned} \tag{31}$$

where $\boldsymbol{\Delta}_{gW}$ is given by (28), $\tilde{\boldsymbol{\Sigma}}_{gm}$ is the reduced form of $\tilde{\boldsymbol{\Sigma}}_g$ given in (19) without the covariances related to z_g in (1). $\tilde{\boldsymbol{\Delta}}_{gm} = \partial \text{vec}(\tilde{\boldsymbol{\Sigma}}_{gm})'/\partial \boldsymbol{\theta}$. $\tilde{\boldsymbol{S}}_{gm}$ is the reduced form of $\tilde{\boldsymbol{S}}_g$ given in (19) without the variate z_g. It can be derived that

$$\tilde{\Sigma}_{gm} = \begin{pmatrix} \Sigma_{gB} + \mu_g\mu_g' & \mu_g \\ \mu_g' & 1 \end{pmatrix}, \quad \tilde{S}_{gm} = \begin{pmatrix} C_g + a_g a_g' & a_g \\ a_g' & 1 \end{pmatrix},$$

$$\begin{aligned} C_g &= \Sigma_{gB} - \Sigma_{gB}(\Sigma_{gB} + \Sigma_{gW})^{-1}\Sigma_{gB}, \\ a_g &= \mu_g + \Sigma_{gB}(\Sigma_{gB} + \Sigma_{gW})^{-1}(y_g - \mu_g), \\ S_{gW} &= y_g y_g' + C_g + a_g a_g' - a_g y_g' - (a_g y_g')', \end{aligned} \tag{32}$$

where μ_g, Σ_{gW}, and Σ_{gB} are given in (8). y_g is the observation from the LMEM (5). The derivatives in (28) help to compute the derivative $\tilde{\Delta}_{gm}$ in (29).

3.2 Asymptotic Properties

Because the estimator $\hat{\theta}$ for the model parameter θ in (1) obtained from the EM algorithm in Sect. 3.1 is an MLE, some general properties such as asymptotic normality apply to the MLE in Sect. 3.1. We can apply the general result on MLE given by Hoadley (1971) to the MLE $\hat{\theta}$. Based on assumptions (A1)–(A5) on (1), the available observations $\{u_g : g = 1, \ldots, G\}$ defined in (9) are independent but not identically distributed. The negative twice of the log-likelihood function from $\{u_g\}$ defined in (9) can be expressed as

$$\begin{aligned} f(\theta) = &\sum_{g=1}^{G} \left\{ (N_g - 1) \log |\Sigma_{gW}(\theta)| + \mathrm{tr}\left[\Sigma_{gW}^{-1}(\theta) T_{gW} \right] \right\} \\ &+ \sum_{g=1}^{G} \left\{ \log |\Omega_g(\theta)| + \mathrm{tr}\left[\Omega_g^{-1}(\theta) T_{gB} \right] \right\}, \end{aligned} \tag{33}$$

where Ω_g is given in (15) and

$$T_{gW} = \frac{1}{N_g - 1} Y_{g0}' \left(I_{N_g} - \frac{1}{N_g} J_{N_g} \right) Y_{g0}, \quad T_{gB} = (c_g - \mu_g)(c_g - \mu_g)', \tag{34}$$

where Y_{g0} is given in (9), and μ_g and c_g are given in (2) and (15), respectively. I_{N_g} is the $N_g \times N_g$ identity matrix and J_{N_g} the $N_g \times N_g$ matrix of ones (all of its elements are "1"). Because the observation vectors $\{u_g\}$ in (9) are independently normally distributed, it can be verified that $\{u_g\}$ satisfy the regularity conditions in *Theorem 2* of Hoadley (1971). Then we have the following theorem.

Theorem 1. *The MLE $\hat{\theta}$ for the model parameter θ from (1) with assumptions (A1)–(A5) is asymptotically normally distributed with*

$$G^{1/2}(\hat{\theta} - \theta) \xrightarrow{\mathcal{D}} N\left(0, 2\Gamma^{-1}(\theta)\right), \quad G \to \infty, \tag{35}$$

where the sign "$\overset{\mathcal{D}}{\rightarrow}$" means "converge in distribution," and the matrix $\boldsymbol{\Gamma}(\boldsymbol{\theta})$ is given by

$$\boldsymbol{\Gamma}(\boldsymbol{\theta}) = \frac{1}{G}\left\{\sum_{g=1}^{G}(N_g - 1)\boldsymbol{\Delta}_{gW}\left(\boldsymbol{\Sigma}_{gW}^{-1} \otimes \boldsymbol{\Sigma}_{gW}^{-1}\right)\boldsymbol{\Delta}'_{gW} + \sum_{g=1}^{G}\left[\boldsymbol{\Delta}_g\left(\boldsymbol{\Omega}_g^{-1} \otimes \boldsymbol{\Omega}_g^{-1}\right)\boldsymbol{\Delta}'_g + 2\boldsymbol{\Delta}_{g\mu}\boldsymbol{\Omega}_g^{-1}\boldsymbol{\Delta}'_{g\mu}\right]\right\}, \tag{36}$$

where $\boldsymbol{\Omega}_g$ and $\boldsymbol{\Delta}_{gW}$ are given in (15) and (22), respectively, and

$$\boldsymbol{\Delta}_g = \frac{\partial(vec\boldsymbol{\Omega}_g)'}{\partial\boldsymbol{\theta}}, \quad \boldsymbol{\Delta}_{g\mu} = \frac{\partial\boldsymbol{\mu}'_g}{\partial\boldsymbol{\theta}}. \tag{37}$$

Theorem 3.1 is a direct result from *Theorem 2* of Hoadley (1971) applied to the independently not identically normally distributed observation vectors $\{\boldsymbol{u}_g\}$ in (9). By Theorem 3.1, the asymptotic standard errors of the components of the MLE $\hat{\boldsymbol{\theta}}$ from the model given by (1) can be approximately computed by the square roots of the corresponding diagonal elements of the asymptotic covariance matrix of $\hat{\boldsymbol{\theta}}$

$$\text{cov}(\hat{\boldsymbol{\theta}}) \approx \frac{2}{G}\left[\boldsymbol{\Gamma}(\hat{\boldsymbol{\theta}})\right]^{-1}. \tag{38}$$

The chi-square statistic for testing goodness-of-fit of formulation (1) with mean and covariance structures (3) is defined as the difference between the model chi-square at the restricted model ((3) with $r < R$, R given by (4)), and the model chi-square at the saturated model ($r = R$) (see, e.g., Lee and Poon, 1998; Liang and Bentler, 2004)

$$\chi^2_{\text{SEM}} = f(\hat{\boldsymbol{\theta}}) - f(\hat{\boldsymbol{\theta}}_s) \overset{\mathcal{D}}{\rightarrow} \chi^2(m), \qquad m = R - r, \tag{39}$$

where $f(\boldsymbol{\theta})$ is the likelihood function given by (31), $\hat{\boldsymbol{\theta}}$ is the MLE of $\boldsymbol{\theta}$ from the restricted model (3) and $\hat{\boldsymbol{\theta}}_s$ the MLE of $\boldsymbol{\theta}_s$ from the saturated model. A smaller χ^2_{SEM}-value implies a larger p-value for the test or better goodness-of-fit of the model. The MLE $\hat{\boldsymbol{\theta}}_s$ from the saturated model for the case of SEM can be obtained by using the updating process given by (26), where the derivative matrices $\boldsymbol{\Delta}_{gW}$ and $\tilde{\boldsymbol{\Delta}}_g$ in (21) and (23) are all constant matrices with elements zeros and ones.

For LMEM (5), we have the following simplified expressions:

(a) The likelihood function (31) reduces to

$$g(\boldsymbol{\theta}) = \sum_{g=1}^{G}\left\{\log|\boldsymbol{\Sigma}_{gB}(\boldsymbol{\theta}) + \boldsymbol{\Sigma}_{gW}(\boldsymbol{\theta})| + \text{tr}\left[\left(\boldsymbol{\Sigma}_{gB}(\boldsymbol{\theta}) + \boldsymbol{\Sigma}_{gW}(\boldsymbol{\theta})\right)^{-1}\boldsymbol{T}_{gB}\right]\right\}, \tag{40}$$

where $\boldsymbol{\Sigma}_{gB}$ and $\boldsymbol{\Sigma}_{gW}$ are given by (8), $\boldsymbol{T}_{gB}$ is given by

$$T_{gB} = (\boldsymbol{y}_g - \boldsymbol{X}'_g\boldsymbol{\beta})(\boldsymbol{y}_g - \boldsymbol{X}'_g\boldsymbol{\beta})' \tag{41}$$

with the observation $\boldsymbol{y}_g$ and the design matrix $\boldsymbol{X}'_g$ given in the LMEM (5).

(b) The asymptotic covariance matrix $\boldsymbol{\Gamma}(\boldsymbol{\theta})$ in (34) for the MLE $\hat{\boldsymbol{\theta}}_L$ (the MLE for $\boldsymbol{\theta}$ in the mean and covariance structures (8)) reduces to

$$\begin{aligned}\boldsymbol{\Phi}(\boldsymbol{\theta}) = \sum_{g=1}^{G} \Big\{ (\boldsymbol{\Delta}_{gB} + \boldsymbol{\Delta}_{gW}) \Big[(\boldsymbol{\Sigma}_{gB} + \boldsymbol{\Sigma}_{gW})^{-1} \otimes (\boldsymbol{\Sigma}_{gB} + \boldsymbol{\Sigma}_{gW})^{-1} \Big] \\ \times (\boldsymbol{\Delta}_{gB} + \boldsymbol{\Delta}_{gW})' + 2\boldsymbol{\Delta}_{g\mu}(\boldsymbol{\Sigma}_{gB} + \boldsymbol{\Sigma}_{gW})^{-1}\boldsymbol{\Delta}'_{g\mu} \Big\},\end{aligned} \tag{42}$$

where $\boldsymbol{\Delta}_{g\mu}$, $\boldsymbol{\Delta}_{gW}$ and $\boldsymbol{\Delta}_{gB}$ are given in (28). The asymptotic standard error of $\hat{\boldsymbol{\theta}}_L$ can be approximately computed by the square roots of the corresponding diagonal elements of the asymptotic covariance matrix of $\hat{\boldsymbol{\theta}}_L$

$$\mathrm{cov}(\hat{\boldsymbol{\theta}}_L) \approx \frac{2}{G}\Big[\boldsymbol{\Phi}(\hat{\boldsymbol{\theta}}_L)\Big]^{-1}. \tag{43}$$

Due to the doubtably estimable problem of the saturated model associated with an LMEM, a model fit test for an LMEM based on the chi-square approach cannot be defined easily and this is beyond the scope of this chapter. The simple expression (38) is useful for model selection related to the likelihood ratio criterion when a number of candidate structured LMEM's are to be compared, see, for example, the AIC and BIC criteria discussed by Kuha (2004).

4 Examples

In this section, we will provide two examples to illustrate the application of the model formulation (1) with the EM algorithm in Sect. 3.

Example 1. (The case of identical within covariance matrices with two pure level-2 observations, see Liang and Bentler, 2004.) The practical two-level data set contains the scores of four tests given to high school students in 1988. There are $N = 5,198$ students nested in $G = 235$ schools. The full data set is available upon request from the authors. We consider the following model (Fig. 1) with only one within covariance matrix. This is a one-factor (FW) model for within student test performance, and a one-factor (FB) model to describe between school differences in student test performance, with FB also predicted by school level variables Z1 and Z2. The intercepts of FB, Z1, and Z2 are given by the paths from the constant 1 (called V999 in the figure). The model parameters are

(a) Within factor loadings: th1 (θ_1), th2 (θ_2), th3 (θ_3).
(b) Within unique variances: th5 (θ_5), th6 (θ_6), th7 (θ_7), th8 (θ_8).
(c) Between factor loadings: ph1 (ϕ_1), ph2 (ϕ_2), ph3 (ϕ_3).
(d) Between unique variances: ph4 (ϕ_4), ph5 (ϕ_5), ph6 (ϕ_6), ph7 (ϕ_7).

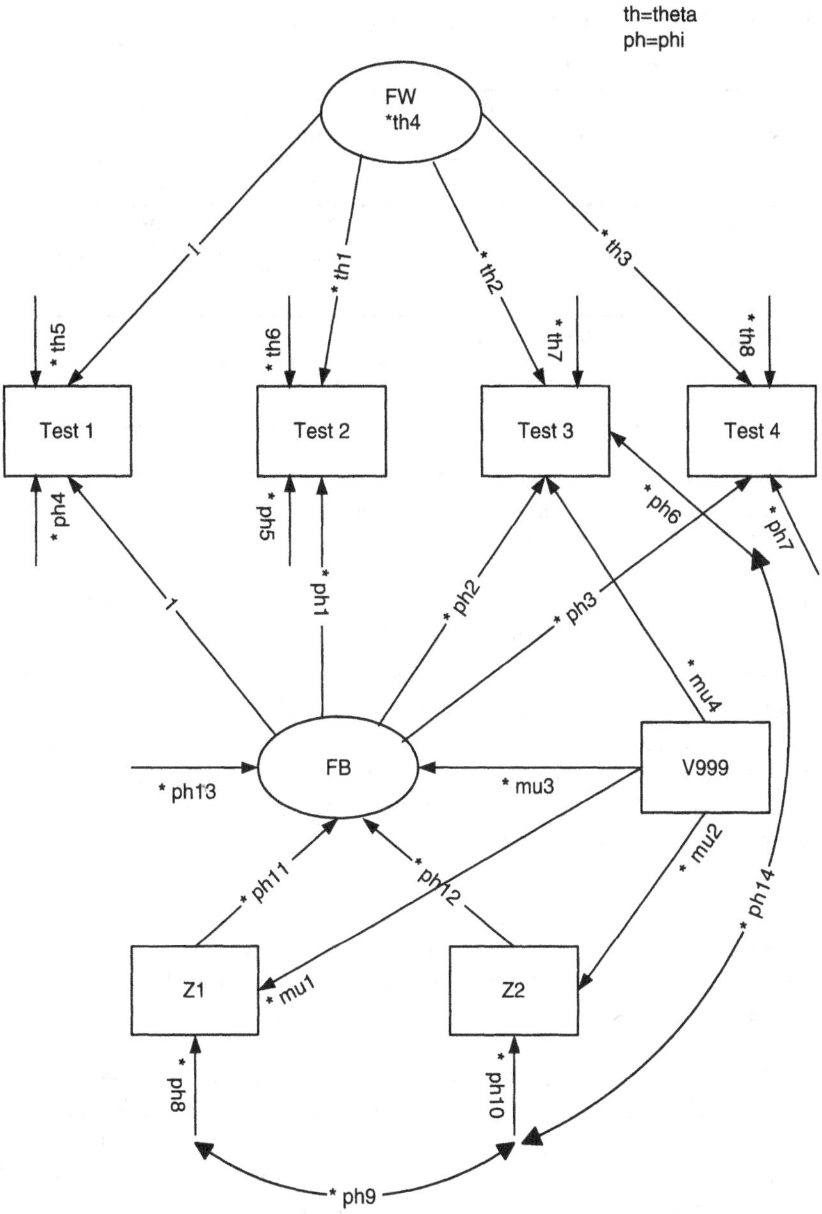

Fig. 1 EQS diagram for the model in Example 1

(e) Coefficients for the between effects of Z1 and Z2 on FB: ph11 (ϕ_{11}), ph12 (ϕ_{12}).
(f) Covariances between residuals: Z1 and Z2, ph9 (ϕ_9), and Z2 and Test 3, ph14 (ϕ_{14}).
(g) Variance of the residual of factor FB: ph13 (ϕ_{13}).
(h) Intercepts: mu1 (μ_1) for Z1, mu2 (μ_2) for Z2, mu3 (μ_3) for FB, and mu4 (μ_4) for Test 3.

Table 1 The ML estimates, standard errors, and chi-square test

	θ_1	θ_2	θ_3	θ_4	θ_5	θ_6	θ_7	θ_8
Estimate	1.046	0.682	1.024	0.751	0.341	0.255	1.348	2.482
SE	0.021	0.022	0.031	0.024	0.015	0.015	0.028	0.053
	ϕ_1	ϕ_2	ϕ_3	ϕ_4	ϕ_5	ϕ_6	ϕ_7	ϕ_8
Estimate	1.052	0.579	1.177	0.014	0.009	0.033	0.051	4.437
SE	0.006	0.043	0.012	0.005	0.004	0.009	0.017	0.409
	ϕ_9	ϕ_{10}	ϕ_{11}	ϕ_{12}	ϕ_{13}	ϕ_{14}		
Estimate	−0.366	0.142	−0.175	−0.025	0.132	−0.023		
SE	0.057	0.013	0.015	0.083	0.017	0.007		
	μ_1		μ_2		μ_3		μ_4	
Estimate	4.630		1.170		3.262		0.728	
SE	0.137		0.025		0.145		0.105	
Test	Chi-square=12.51, df=11, p-Value=0.33							

Therefore, we have a total of 26 ($r = 26$) model parameters to be estimated under the null hypothesis that the model specified by Fig. 1 is true.

The EM algorithm for identical within covariance matrices has been coded into EQS 6.1 (Bentler, 2006). We ran the EQS program and obtained the parameter estimates in Table 1. The p-value=0.33 of the model chi-square indicates a good model fit. That is, the restricted model given by Fig. 1 is suitable for representing the data. The EQS program for running the model specified by Fig. 1, the formulae for computing the standard error (SE) and the model chi-square in Table 1 are given in Liang and Bentler (2004).

Example 2. (The case of nonidentical within covariance matrices with two pure level-2 observations.) The data set is selected from the full data set in Example 1. We set up a model with two different within covariance matrices. For the purpose of illustration, it is assumed that there are 20 schools in which the students are only given three tests and 215 schools in which the students are given four tests. The model for the 20 schools with three tests is given by Fig. 2 and the model for the 215 schools with four tests is given by Fig. 3. Some parameter constraints are assumed to hold between the model in Fig. 2 and the model in Fig. 3. For example, the set of all parameters in Fig. 2 is a subset of the parameters in Fig. 3. That is, all parameters in Fig. 2 are constrained to be equal to the corresponding parameters in Fig. 3. The SEM version of the algorithm in Sect. 3 has been coded into the multilevel option in the current EQS version (EQS 6.1). For readers' easy application of the EM algorithm in Sect. 3 in two-level SEM with dimensional heterogeneity, we provide the EQS setup for running the model specified by Figs. 2 and 3 in Appendix. The EQS output for the parameter estimates, their standard errors and model chi-square is summarized in Table 2. The p-value=0.15 of the model chi-square indicates that the restricted model specified by Figs. 2 and 3 is suitable for representing the selected data.

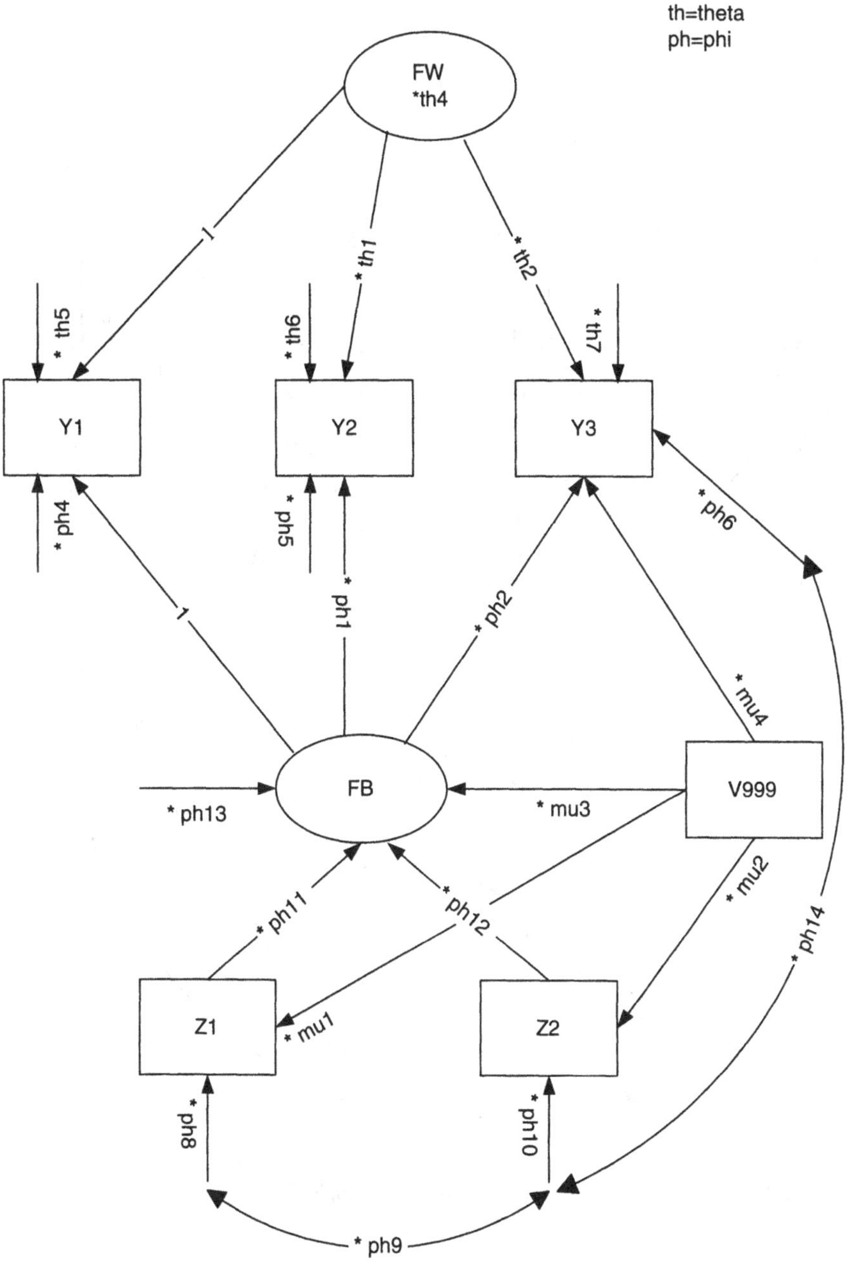

Fig. 2 EQS diagram for model 1 with 20 schools in Example 2

In the above analysis for Example 2, it could be argued that the model (Fig. 2) for the first 20 schools with only three tests given can be considered as a missing data pattern for the model (Fig. 3) for the last 215 schools with four tests given. In the case of missing data pattern, there is only one saturated model (associated with

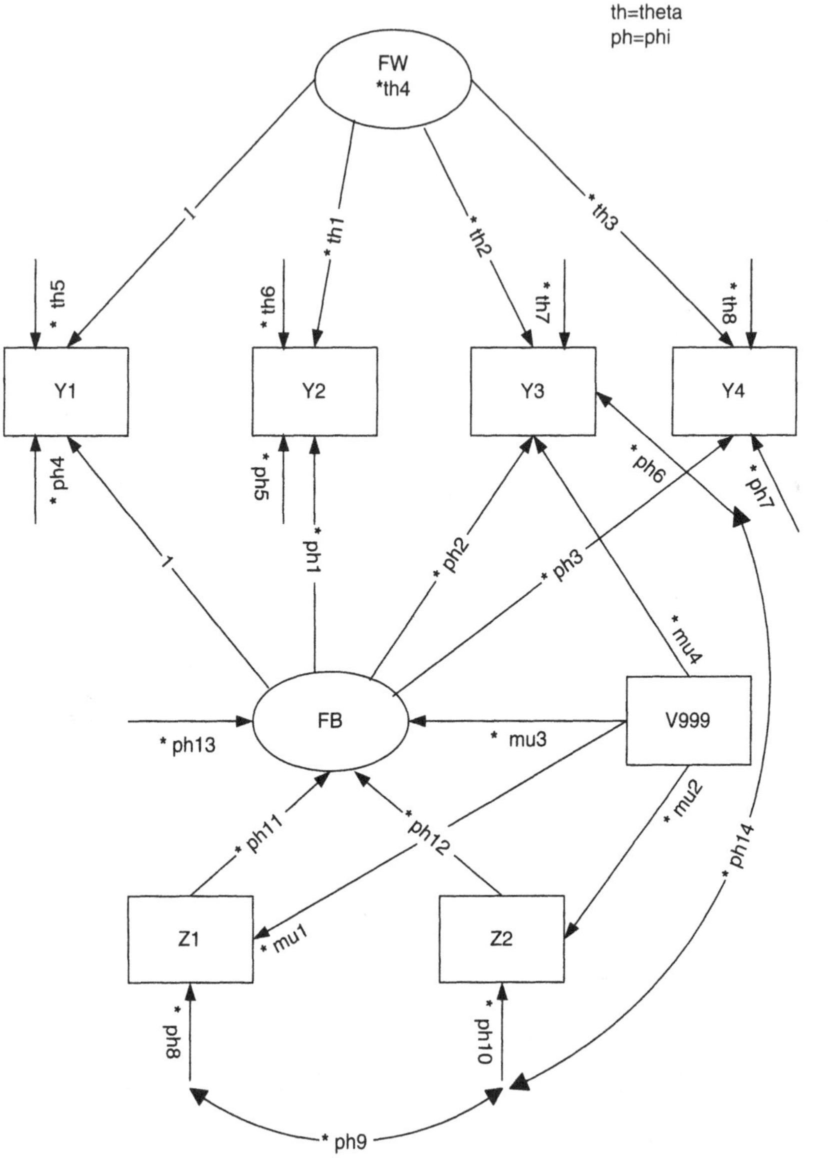

Fig. 3 EQS diagram for model 2 with 215 schools in Example 2

Fig. 3) with the number of saturated model parameters (computed from (4) with $G = 1$, $p = 4$ and $q = 2$).

$$R = 4 + 2 + 4(4 + 1)/2 + (4 + 2)(4 + 2 + 1)/2 = 37.$$

The number of structured model parameters $r = 26$ from Fig. 3. Then the number of degrees of freedom is df=37−26=11 instead of df=37 as reported in Table 2. This can be interpreted by the definition of a saturated model. When using the missing

Table 2 The ML estimates, standard errors, and chi-square test

Parameter	θ_1	θ_2	θ_3	θ_4	θ_5	θ_6	θ_7	θ_8
MLE	1.049	0.683	1.019	0.749	0.343	0.253	1.348	2.478
SE	0.022	0.022	0.032	0.025	0.015	0.015	0.028	0.055
Parameter	ϕ_1	ϕ_2	ϕ_3	ϕ_4	ϕ_5	ϕ_6	ϕ_7	ϕ_8
MLE	1.052	0.579	1.180	0.013	0.010	0.036	0.056	4.437
SE	0.006	0.043	0.013	0.005	0.005	0.009	0.018	0.409
Parameter	ϕ_9	ϕ_{10}	ϕ_{11}	ϕ_{12}	ϕ_{13}	ϕ_{14}		
MLE	−0.366	0.143	−0.174	−0.023	0.131	−0.026		
SE	0.057	0.013	0.015	0.082	0.016	0.007		
Parameter	μ_1		μ_2		μ_3		μ_4	
MLE	4.630		1.170		3.259		0.729	
SE	0.137		0.025		0.145		0.107	
Model chi-square=46.07, df=37, p-Value=0.15								

data pattern, the equal-parameter constraint is imposed on both models for the first 20 schools (Fig. 2) and for the last 215 schools (Fig. 3). So the saturated model under the missing data pattern is not completely saturated. The number of degrees of freedom df=37 from the EQS output in Table 2 is derived from (4) by taking complete saturated models for the first 20 schools (Fig. 2) and for the last 215 schools (Fig. 3), respectively. This is actually equivalent to the multiple-group multilevel analysis. The saturated model associated with this analysis should not include any parametric constraints across the groups. So the number of degrees of freedom for the chi-square test should be computed from (4) with $G = 2$, $p_1 = 3$, $q_1 = 2$, $p_2 = 4$, $q_2 = 2$, and

$$\begin{aligned} R = {} & [3 + 2 + 3(3 + 1)/2 + (3 + 2)(3 + 2 + 1)/2] \\ & + [4 + 2 + 4(4 + 1)/2 + (4 + 2)(4 + 2 + 1)/2] = 63. \end{aligned} \tag{44}$$

The number of structured model parameters $r = 26$ from Fig. 3 due to the equality constraints on the parameters from both Figs. 2 and 3. Therefore the number of degrees of freedom for the chi-square test is df=$R - r = 63 - 26 = 37$ as reported from the EQS output in Table 2.

When there are more than two different within covariance matrices and more than one between covariance matrix, as for the case of LMEM, optimization of the E-step function (18) is carried out by the M-step iteration (26). The multilevel option in the current EQS program can provide the MLE of model parameter for the SEM version of the algorithm. When the EM algorithm in Sect. 3 is applied to estimating model parameters from an LMEM, the computation of the derivatives of the different within and between covariance matrices versus the unknown model parameter can be greatly simplified by (28)–(30). Then the M-step iteration (26) for optimizing the E-step function (18) will be much faster than the general case of many completely different within and between covariance matrices.

5 Goodness-of-Fit and Related Issues

In the previous sections, we introduced a common model formulation for analysis of two-level SEM and LMEM and illustrated its applications by practical data sets. Statistical models are usually set up for specifically collected data sets. It is usually necessary to test whether a proposed model is suitable for a given data set. This is an issue of testing goodness-of-fit and model selection. In covariance structure analysis, testing goodness-of-fit is equivalent to testing a restricted model (the null hypothesis) versus a more general (less restrictive) model (the alternative hypothesis). In analysis of conventional (one-level) SEM, there are quite a few methods for testing SEM and related discussions as mentioned in Sect. 1. In analysis of two-level SEM, only a few existing methods for testing goodness-of-fit are available, see Bentler et al. (2005) and references therein.

In addition to the usual chi-square statistic and some modified chi-square statistics as summarized in Bentler et al. (2005) for model evaluation in two-level SEM, there are some other model selection techniques for model assessment, model improvement, model modification, and model comparison, see, for example, Akaike (1987), Bozdogan (1987), Browne and Cudeck (1989), Burnham and Anderson (2004), Kuha (2004), Sörbom (1989), Yuan (2005), and Yuan et al. (2002). Because many model selection techniques such as the AIC and BIC (Kuha, 2004) require computation of the value of the log-likelihood function at the MLE of the model parameters, a simple expression for the log-likelihood function and an easily-implemented algorithm for obtaining the MLE of the parameters are essential to applying various model selection techniques in practice. Formulation (1) with the assumptions (A1)–(A5) has a simple expression (31) for the log-likelihood function of two-level SEM. By using the multilevel option in EQS to run an EQS program under the model formulation (1), we can easily obtain the MLE of the model parameters under model formulation (1) for two-level SEM. Therefore, the above-mentioned existing model selection techniques could be applied to model formulation (1) to help obtain a suitable model for a given two-level data set. The simplicity of the model formulation (1) may also provide a convenient way to implement Bayesian model selection if some prior information on the model parameters is available. Based on the simple log-likelihood function (31) for formulation (1), under the normal assumption, the Bayesian model selection method in Song and Lee (2002) could be employed to select a suitable model if there are several candidate models with the same formulation (1). Further discussions on Bayesian model selection can be found in Gelman and Rubin (1995), Lee and Song (2003), and Raftery (1995a,b).

In this chapter, we proposed a unified model formulation for analysis of two-level SEM and LMEM with the focus on parameter estimation. Due to the difficulty of specifying the saturated models associated with many LMEM's, the same chi-square-type statistic as for two-level SEM is not suitable for testing goodness-of-fit for an LMEM when its associated saturated model is not estimable. It is usually a complicated task to verify whether the saturated model associated with a structured LMEM is estimable. So we were not able to provide an illustration on LMEM with practical data for which an estimable saturated model can be set up. By resorting to the simple likelihood function (38) and the proposed EM algorithm in Sect. 3 to obtain the MLE's for a number of candidate structured LMEM's for a given data set, one will be able to use some existing criteria related to the likelihood function (such as the AIC and BIC (Kuha, 2004)) for model selection and model comparison. Therefore, the unified estimation approach to two-level SEM and LMEM in this chapter could provide a convenient way to apply existing methodologies in model selection to the unified model formulation for two-level SEM and LMEM.

It is pointed out that the unified estimation approach to two-level SEM and LMEM in this chapter is strictly based on the normal assumptions as specified by (A1)–(A4) in Sect. 2. Some robust procedures against violation of the normality assumption in SEM were discussed by Yuan and Bentler (2007b). Since an LMEM can be fit into the same formulation (1) as for two-level SEM for parameter estimation, we could expect some kind of robustness of the estimation method in this chapter, for example, the asymptotic robustness of standard errors discussed by Yuan and Bentler (2006). Further research is necessary to bridge the relation between SEM and LMEM in view of estimation methods, goodness-of-fit statistics, and robust procedures.

Appendix. EQS Input Program for the Model in Example 2

```
/TITLE
First set of schools – only 3 variables
WITHIN MODEL FIRST
/SPECIFICATION
data='school1.dat'; case =447; variable=7; method=ml;
matrix=raw; GROUP=4; analysis=covariance; MULTILEVEL=ML; cluster=V5;
/LABELS
V1 = Y6; V2 = Y7; V3 = Y8; V4 = Y9; V5 = SCHOOL; V6 = X3; V7 = X4;
F1 = FW;
/EQUATIONS
Y6=1FW+E1;
Y7=*FW+E2;
Y8=*FW+E3;
!Y9=*FW+E4;
/VARIANCES
FW=*;
E1-E3=*;
/END
/TITLE
BETWEEN MODEL
```

(To be continued)

(Continued)

```
/LABELS
V1 = Y6; V2 = Y7; V3 = Y8; V4 = Y9; V5 = SCHOOL; V6 = X3; V7 = X4;
F1 = FB;
/EQUATIONS
Y6=1FB+E1;
Y7=*FB+E2;
Y8=*V999+*FB+E3;
!Y9=*FB+E4;
FB=*V999+*X3+*X4+D1;
X3=*V999+E20;
X4=*V999+E21;
/VARIANCES
E1-E3=*;
D1=*;
E20-E21=*;
/COVARIANCES
E20,E21=*;
E3,E21=*;
/END
/TITLE
Second set of schools – all four variables
WITHIN MODEL FIRST
/SPECIFICATION
data='school2.dat'; case =4751; variable=7; method=ml;
matrix=raw; analysis=covariance; MULTILEVEL=ML; cluster=V5;
/LABELS
V1 = Y6; V2 = Y7; V3 = Y8; V4 = Y9; V5 = SCHOOL; V6 = X3; V7 = X4;
F1 = FW;
/EQUATIONS
Y6=1FW+E1;
Y7=*FW+E2;
Y8=*FW+E3;
Y9=*FW+E4;
/VARIANCES
FW=*;
E1-E4=*;
/END
/TITLE
BETWEEN MODEL
/LABELS
```

(To be continued)

(Continued)

```
V1 = Y6; V2 = Y7; V3 = Y8; V4 = Y9; V5 = SCHOOL; V6 = X3; V7 = X4;
F1 = FB;
/EQUATIONS
Y6=1FB+E1;
Y7=*FB+E2;
Y8=*V999+*FB+E3;
Y9=*FB+E4;
FB=*V999+*X3+*X4+D1;
X3=*V999+E20;
X4=*V999+E21;
/VARIANCES
E1-E4=*;
D1=*;
E20-E21=*;
/COVARIANCES
E20,E21=*;
E3,E21=*;
/CONSTRAINTS
(1,F1,F1)-(3,F1,F1)=0;
(1,E1,E1)-(3,E1,E1)=0;
(2,E1,E1)-(4,E1,E1)=0;
(1,E2,E2)-(3,E2,E2)=0;
(2,E2,E2)-(4,E2,E2)=0;
(1,E3,E3)-(3,E3,E3)=0;
(2,E3,E3)-(4,E3,E3)=0;
(2,E20,E20)-(4,E20,E20)=0;
(2,E21,E3)-(4,E21,E3)=0;
(2,E21,E20)-(4,E21,E20)=0;
(2,E21,E21)-(4,E21,E21)=0;
(2,D1,D1)-(4,D1,D1)=0;
(2,V3,V999)-(4,V3,V999)=0;
(2,V6,V999)-(4,V6,V999)=0;
(2,V7,V999)-(4,V7,V999)=0;
(1,V2,F1)-(3,V2,F1)=0;
(1,V3,F1)-(3,V3,F1)=0;
(2,F1,V999)-(4,F1,V999)=0;
(2,V2,F1)-(4,V2,F1)=0;
(2,V3,F1)-(4,V3,F1)=0;
(2,F1,V6)-(4,F1,V6)=0;
(2,F1,V7)-(4,F1,V7)=0;
/tech
itr=200; con=.000001;
/END
```

Acknowledgement This work was supported by National Institute on Drug Abuse Grants DA01070, DA00017, and the University of New Haven 2006 and 2007 Summer Faculty Fellowships.

References

Akaike, H. (1987). Factor analysis and AIC. *Psychometrika*, 52, 317–332

Bauer, D. J. (2003). Estimating multilevel linear models as structural models. *Journal of Educational and Behavioral Statistics*, 28, 135–167

Bentler, P. M. (2006). *EQS 6 Structural Equations Program Manual*. Encino, CA: Multivariate Software (www.mvsoft.com)

Bentler, P. M. and Liang, J. (2003). Two-level mean and covariance structures: maximum likelihood via an EM algorithm. In S. Reise and N. Duan (Eds.), *Multilevel Modeling: Methodological Advances, Issues, and Applications* (pp. 53–70). Mahwah, NJ: Lawrence Erlbaum Associates

Bentler, P. M., Liang, J. and Yuan, K.-H. (2005). Some recent advances in two-level structural equation models: estimation, testing and robustness. In J. Fan and G. Li (Eds.), *Contemporary Multivariate Analysis and Experimental Design* (pp. 211–232). NJ: World Scientific Publisher

Bozdogan, H. (1987). Model selection and Akaike's information criterion (AIC): the central theory and its analytical extensions. *Psychometrika*, 52, 345–370

Browne, M. W. and Cudeck, R. (1989). Single sample cross-validation indices for covariance structures. *Multivariate Behavioral Research*, 24, 445–455

Burnham, K. P. and Anderson, D. R. (2004). Multimodal inference: understanding AIC and BIC in model selection. *Sociological Methods & Research*, 33, 261–304

Cudeck, R. and Browne, M. W. (1983). Cross-validation of covariance structures. *Multivariate Behavioral Research*, 18, 147–157

Curran, P. J. (2003). Have multilevel models been structural equation models all along? *Multivariate Behavioral Research*, 38, 529–569

Dempster, A. P., Laird, N. M. and Rubin, D. B. (1977). Maximum likelihood from incomplete data via EM algorithm (with discussion). *Journal of the Royal Statistical Society (Series B)*, 39, 1–38

du Toit, M. and du Toit, S. (2001). *Interactive LISREL: User's Guide*. Chicago: Scientific Software International, Inc

Gelman, A. and Rubin, D. B. (1995). Avoiding model selection in Bayesian social research. In A. E. Raftery (Ed.), *Sociological Methodology* (pp. 165–173). Oxford, UK: Blackwell

Hoadley, B. (1971). Asymptotic properties of maximum likelihood estimators for the independent not identically distributed case. *The Annals of Mathematical Statistics*, 42, 1977–1991

Kuha, J. (2004). AIC and BIC: comparison of assumptions and performance. *Sociological Methods & Research*, 33, 188–229

Lange, K. (1995a). A gradient algorithm locally equivalent to the EM algorithm. *Journal of the Royal Statistical Society (Series B)*, 57, 425–437

Lange, K. (1995b). A Quasi-Newton acceleration of the EM algorithm. *Statistica Sinica*, 5, 1–18

Lee, S. Y. (1990). Multilevel analysis of structural equation models. *Biometrika*, 77, 763–772

Lee, S. Y. and Poon, W. Y. (1998). Analysis of two-level structural equation models via EM type algorithms. *Statistica Sinica*, 8, 749–766

Lee, S.-Y. and Song, X.-Y. (2003). Bayesian model selection for mixtures of structural equation models with an unknown number of components. *British Journal of Mathematical and Statistical Psychology*, 56, 145–165

Liang, J. and Bentler, P. M. (2004). An EM algorithm for fitting two-level structural equation models. *Psychometrika*, 69, 101–122

McDonald, R. P. and Goldstein, H. (1989). Balanced versus unbalanced designs for linear structural relations in two-level data. *British Journal of Mathematical and Statistical Psychology*, 42, 215–232

McLachlan, G. J. and Krishnan, T. (1997). *The EM Algorithm and Extensions*. New York: Wiley

Mehta, P. D. and Neale, M. C. (2005). People are variables too: multilevel structural equations modeling. *Psychological Methods*, 10, 259–284

Meng, X.-L. (1998). Fast EM-type implementations for mixed effects models. *Journal of the Royal Statistical Society (Series B)*, 60, 559–578

Muthén, B. O. (1989). Latent variable modeling in heterogeneous populations. *Psychometrika*, 54, 557–585

Muthén, B. (1997). Latent variable modeling with longitudinal and multilevel data. In A. Raftery (Ed.), *Sociological methodology* (pp. 453–480). Boston: Blackwell Publishers

Muthén, L. K. and Muthén, B. O. (2004). *Mplus User's Guide*. Los Angeles: Muthén & Muthén

Raftery, A. E. (1995a). Bayesian model selection in social research. In A. E. Raftery (Ed.), *Sociological Methodology* (pp. 111–164). Oxford, UK: Blackwell

Raftery, A. E. (1995b). Rejoinder: model selection is unavoidable in social research. In A. E. Raftery (Ed.), *Sociological Methodology* (pp. 185–195). Oxford, UK: Blackwell

Raudenbush, S. W. (1995). Maximum likelihood estimation for unbalanced multilevel covariance structure models via the EM algorithm. *British Journal of Mathematical and Statistical Psychology*, 48, 359–370

Rovine, M. J. and Molenaar, P. C. (2000). A structural modeling approach to a multilevel random coefficients model. *Multivariate Behavioral Research*, 35, 51–88

Skrondal, A. and Rabe-Hesketh, S. (2004). *Generalized Latent Variable Modeling: Multilevel, Longitudinal and Structural Equation Models*. Boca Raton, FL: Chapman and Hill

Song, X.-Y. and Lee, S.-Y. (2002). A Bayesian model selection method with applications. *Computational Statistics & Data Analysis*, 40, 539–557

Sörbom, D. (1989). Model modification. *Psychometrika*, 54, 371–384

Yuan, K.-H. (2005). Fit indices versus test statistics. *Multivariate Behavioral Research*, 40, 115–148

Yuan, K.-H. and Bentler, P. M. (2003). Eight test statistics for multilevel structural equation models. *Computational Statistics & Data Analysis*, 44, 89–107

Yuan, K.-H. and Bentler, P. M. (2006). Asymptotic robustness of standard errors in multilevel structural equation models. *Journal of Multivariate Analysis*, 97, 1121–1141

Yuan, K.-H. and Bentler, P. M. (2007a). Structural equation modeling. In C. R. Rao and S. Sinharay (Eds.), *Handbook of Statistics 26: Psychometrics* (pp. 297–358). Amsterdam: North-Holland

Yuan, K.-H. and Bentler, P. M. (2007b). Robust procedures in structural equation modeling. In S.-Y. Lee (Ed.), *Handbook of Latent Variable and Related Models* (pp. 367–397). Amsterdam: Elsevier

Yuan, K.-H., Marshall, L. L. and Weston, R. (2002). Cross-validation by downweighting influential cases in structural equation modeling. *British Journal of Mathematical and Statistical Psychology*, 55, 125–143

Bayesian Model Comparison of Structural Equation Models

Sik-Yum Lee and Xin-Yuan Song

1 Introduction

Structural equation modeling is a multivariate method for establishing meaningful models to investigate the relationships of some latent (causal) and manifest (control) variables with other variables. In the past quarter of a century, it has drawn a great deal of attention in psychometrics and sociometrics, both in terms of theoretical developments and practical applications (see Bentler and Wu, 2002; Bollen, 1989; Jöreskog and Sörbom, 1996; Lee, 2007). Although not to the extent that they have been used in behavioral, educational, and social sciences, structural equation models (SEMs) have been widely used in public health, biological, and medical research (see Bentler and Stein, 1992; Liu et al., 2005; Pugesek et al., 2003 and references therein). A review of the basic SEM with applicants to environmental epidemiology has been given by Sanchez et al. (2005).

One important statistical inference beyond estimation in SEMs is related to model comparison or testing various hypotheses about the model. In the field of structural equation modeling, a classical approach in hypothesis testing is to use the significance tests on the basis of p-values that are determined by some asymptotic distributions of the test statistics. It is well-known that the p-value of a significance test in hypothesis testing is a measure of evidence against the null model, not a mean of supporting/proving the model. For complex SEMs, the asymptotic distributions of the test statistics are usually unknown and hard to derive. Moreover, the significance tests cannot be applied to test nonnested hypotheses or to compare nonnested models. Various descriptive fit indices (Bentler and Bonett, 1980; Kim, 2005) have been proposed for assessing goodness-of-fit of a single hypothesized model. However, it is not clear how to apply these fit indices for model comparison of complex SEMs.

S.-Y. Lee and X.-Y. Song
Department of Statistics, Chinese University of Hong Kong, Shatin, N.T., Hong Kong
sylee@sparc2.sta.cuhk.edu.hk, xysong@sparc2.sta.cuhk.edu.hk

D. B. Dunson (ed.) *Random Effect and Latent Variable Model Selection*,
DOI: 10.1007/978-0-387-76721-5,

The main objective of this chapter is to introduce a well-known model comparison statistic, namely the Bayes factor (see Kass and Raftery, 1995), which does not have the above mentioned problems. Bayes factor is associated with the Bayesian approach, which has been widely used in analyzing various statistical models and data (see, e.g., Austin and Escobar, 2005; Congdon, 2005; Dunson, 2005; Dunson and Herring, 2005; among others), and SEMs (see, e.g., Lee, 2007; Lee and Song, 2003; Palomo et al., 2007; Song and Lee, 2004; among others). It is well-known that the Bayesian approach has the following advantages: (a) More precise estimates of the parameters can be obtained with good prior information. (b) As the sampling-based Bayesian methods do not rely on asymptotic theory, it gives more reliable statistical results under situations with small sample sizes (see, Dunson, 2000; Lee and Song, 2004a). (c) It has the similar optimal asymptotic properties as the maximum likelihood approach.

The basic idea of the Bayes factor is simple; it is defined as a ratio of the marginal densities associated with the competing models. In general, as the marginal densities involve intractable high-dimensional integrals, computation of Bayes factor is challenging and has received much attention in the literature; see, for example, the rough approximation of the Bayes factor via the Schwarz criterion (or Bayesian Information criterion, BIC); the Barlett adjustment of the Laplace approximation (Diciccio et al., 1997), importance sampling and bridge sampling (see Meng and Wong, 1996; Diciccio et al., 1997), reciprocal importance sampling (Gelfand and Dey, 1994), method using MCMC outputs (Chib, 1995; Chib and Jeliazkov, 2001); and path sampling (Gelman and Meng, 1998). Inspired by its simplicity and its successful applications in latent variable models and SEMs (Song and Lee, 2006a,b), path sampling will be used for computing the Bayes factor in this chapter.

In Sect. 2, we introduce the Bayes factor to compare competing SEMs for a given data set, and roughly discuss some other alternatives. The computation of the Bayes factor through path sampling is addressed in Sect. 3. To illustrate the methodology, model comparisons in the context of nonlinear SEM are given in Sect. 4, with results of a simulation study. Moreover, Sect. 5 presents an application of Bayes factor to an integrated SEM, together with a real application in relation to a hierarchical data set in education. Section 6 presents a discussion.

2 Bayes Factor and other Model Comparison Statistics

2.1 Bayes Factor

Bayes factor (Kass and Raftery, 1995) is a well-known statistic in Bayesian hypothesis testing and model comparison, and it still receives a great deal of attention in recent Bayesian literature. Suppose that the observed data $\mathbf{Y}$ with a sample size n have arisen under one of the two competing models M_1 and M_0 according to probability densities $p(\mathbf{Y}|M_1)$ and $p(\mathbf{Y}|M_0)$, respectively. For $k = 0, 1$, let $p(M_k)$

be the prior probability and $p(M_k|\mathbf{Y})$ be the posterior probability. From the Bayes theorem, we obtain

$$\frac{p(M_1|\mathbf{Y})}{p(M_0|\mathbf{Y})} = \frac{p(\mathbf{Y}|M_1)p(M_1)}{p(\mathbf{Y}|M_0)p(M_0)}.$$

Assuming $p(M_1) = p(M_0)$, the Bayes factor for evaluating M_1 against M_0 is defined as

$$B_{10} = \frac{p(\mathbf{Y}|M_1)}{p(\mathbf{Y}|M_0)}.$$

Hence, the Bayes factor is a summary of the evidence provided by the data in favor of M_1 as opposed to M_0, and it measures how well M_1 predicts the data relative to M_0. From Kass and Raftery (1995), it is useful to consider twice the natural logarithm of the Bayes factor and interpret the resulting statistic based on the following table:

B_{10}	$2 \log B_{10}$	Evidence against M_0
<1	<0	Negative (support M_0)
1–3	0–2	Not worth more than a bare mention
3–20	2–6	Positive (support M_1)
20–150	6–10	Strong
>150	>10	Decisive

(1)

The interpretation given in (1) is a suggestion and it is not necessary to regard it as a strict rule, and selection depends on one's preference in the substantive situation. Similarly in frequentist hypothesis testing, one may take the type I error to be 0.05 or 0.10, and the choice is decided with other factors in the substantive situation. See Garcia-Donato and Chan (2005) for more technical treatment on calibrating the Bayes factor.

The prior distributions of the parameters are involved in the Bayes factor, see (2) below. As pointed out by Kass and Raftery (1995), noninformative priors should not be used. In most Bayesian analyses of SEMs, the proper conjugate type prior distributions that involve prior inputs of the hyper-parameter values have been used. For situations where we have good prior information, for example, from analysis of closely related data, or knowledge of experts, subjective hyper-parameter values should be used. In other situations, ideas from data and/or information from various sources can be used.

In general, the marginal densities $p(\mathbf{Y}|M_k)$, $k = 0, 1$, are obtained by integrating over the parameter space, that is,

$$p(\mathbf{Y}|M_k) = \int p(\mathbf{Y}|\boldsymbol{\theta}_k, M_k)p(\boldsymbol{\theta}_k|M_k)\,\mathrm{d}\boldsymbol{\theta}_k, \tag{2}$$

where $\boldsymbol{\theta}_k$ is the parameter vector in M_k, $p(\boldsymbol{\theta}_k|M_k)$ is its prior density, and $p(\mathbf{Y}|\boldsymbol{\theta}_k, M_k)$ is the probability density of $\mathbf{Y}$ given $\boldsymbol{\theta}_k$. In this chapter, we use $\boldsymbol{\theta}$ to represent the parameter vector in general, and it may stand for $\boldsymbol{\theta}_0$, $\boldsymbol{\theta}_1$ or the parameter vector in the linked model (see the following sections) according to the context. The dimension of the above integral is equal to the dimension of $\boldsymbol{\theta}_k$. Very often, it is very difficult to obtain B_{10} analytically, and various analytic and numerical approximations have been proposed in the literature. In the next section, the procedure based on idea of the path sampling (Gelman and Meng, 1998) is proposed to compute the Bayes factor. The path sampling, which is a generalization of the importance sampling and bridge sampling (Meng and Wong, 1996), has several nice features. Its implementation is simple. In general, as pointed out by Gelman and Meng (1998), we can always construct a continuous path to link two competing models with the same support. Hence, the method can be applied to a wide variety of problems. Unlike some methods in estimating the marginal likelihood via posterior simulation, it does not require to estimate the location and/or scale parameters in the posterior. Distinct from most existing approaches, the prior density is not directly involved in the evaluation. Finally, the logarithm scale of Bayes factor is computed, which is generally more stable than the ratio scale.

2.2 Other Alternatives

A simple approximation of $2\log B_{10}$ is the following Schwarz criterion S^*(Schwarz, 1978):

$$2\log B_{10} \cong 2S^* = 2\left\{\log p(\mathbf{Y}|\tilde{\boldsymbol{\theta}}_1, M_1) - \log p(\mathbf{Y}|\tilde{\boldsymbol{\theta}}_0, M_0)\right\} - (d_1 - d_0)\log n, \quad (3)$$

where $\tilde{\boldsymbol{\theta}}_1$ and $\tilde{\boldsymbol{\theta}}_0$ are maximum likelihood (ML) estimates of $\boldsymbol{\theta}_1$ and $\boldsymbol{\theta}_0$ under M_1 and M_0, respectively; d_1 and d_0 are the dimensions of $\boldsymbol{\theta}_1$ and $\boldsymbol{\theta}_0$; and n is the sample size. Minus $2S^*$ is the following Bayesian Information Criterion (BIC) for comparing M_1 and M_0:

$$\mathrm{BIC}_{10} = -2S^* \cong -2\log B_{10} = 2\log B_{01}. \quad (4)$$

The interpretation of BIC_{10} can be based on (1). For each M_k, $k = 0, 1$, we define

$$\mathrm{BIC}_k = -2\log p(\mathbf{Y}|\tilde{\boldsymbol{\theta}}_k, M_k) + d_k \log n. \quad (5)$$

Then $2\log B_{10} = \mathrm{BIC}_0 - \mathrm{BIC}_1$. Hence, it follows that the model M_k with the smaller BIC_k value is selected.

As n tends to infinity, it has been shown (Schwarz, 1978) that

$$\frac{S^* - \log B_{10}}{\log B_{10}} \to 0,$$

thus S^* may be viewed as an approximation to $\log B_{10}$. As this approximation is of order $O(1)$, S^* does not give the exact $\log B_{10}$ even for large samples. However, since the interpretation is on the natural logarithm scale, it provides a reasonable indication of evidence. As pointed out by Kass and Raftery (1995), it can be used for scientific reporting as long as the number of degrees of freedom $(d_1 - d_0)$ involved in the comparison is small relative to the sample size n. The BIC is appealing in that it is relatively simple and can be applied even when the priors $p(\boldsymbol{\theta}_0|M_k)$ $(k = 1, 2)$ are hard to set precisely. The ML estimates of $\boldsymbol{\theta}_1$ and $\boldsymbol{\theta}_0$ are involved in the computation of BIC. In practice, since the Bayesian estimates and the ML estimates are close to each other, it can be used for computing the BIC. The order of approximation is not changed, and the BIC obtained can be interpreted using the criterion given in (1). See Raftery (1993) for an application of BIC to the standard LISREL model that is based on the normal assumption and a linear structural equation. Under this simple case, the computation of the observed data logarithm likelihood $\log p(\mathbf{Y}|\tilde{\boldsymbol{\theta}}_k, M_k)$ is straight forward. However, for some complex SEMs, evaluation of the observed data logarithm likelihood may be difficult.

The Akaike Information Criterion (AIC) associated with a competing model M_k is given by

$$\mathrm{AIC}_k = -2\log p(\mathbf{Y}|\tilde{\boldsymbol{\theta}}_k, M_k) + 2d_k, \tag{6}$$

which does not involve the sample size n. The interpretation of AIC_k is similar to BIC_k. That is, M_k is selected if its AIC_k is smaller. Comparing (5) with (6), we see that BIC tends to favor simpler models than those selected by AIC.

Another goodness-of-fit or model comparison statistic that takes into account the number of unknown parameters in the model is the Deviance Information Criterion (DIC), see Spiegelhalter et al. (2002). This statistic is intended as a generalization of AIC. Under a competing model M_k with a vector of unknown parameter $\boldsymbol{\theta}_k$ of dimension d_k, let $\{\boldsymbol{\theta}_k^{(j)} : j = 1, \ldots, J\}$ be a sample of observations simulated from the posterior distribution. We define

$$\mathrm{DIC}_k = -\frac{2}{J}\sum_{j=1}^{J} \log p\left(\mathbf{Y}|\boldsymbol{\theta}_k^{(j)}, M_k\right) + 2d_k. \tag{7}$$

In model comparison, the model with the smaller DIC value is selected.

In analyzing a hypothesized model, WinBUGS (Spiegelhalter et al., 2003) produces a DIC value that can be used for model comparison. However, as pointed out in the WinBUGS User Manual (Spiegelhalter et al., 2003), in practical application of DIC, it is important to note the following: (a) If the difference in DIC is small, for example, less than five, and the models make very different inferences, then just reporting the model with the lowest DIC could be misleading. (b) DIC can be applied to nonnested models. Similar to the Bayes factor, BIC, and AIC, DIC gives clear conclusion to support the null hypothesis or the alternative hypothesis. (c) DIC assumes the posterior mean to be a good estimate of the parameter. There are circumstances, such as mixture models, in which WinBUGS will not give the DIC values. Because of (c) and because the Bayes factor is the more common statistic for model comparison, we focus on Bayes factor in this chapter.

3 Computation of Bayes Factor through Path Sampling

In general, let $\mathbf{Y}$ be the matrix of observed data, $\mathbf{L}$ be the matrix of latent data, and $\boldsymbol{\Omega}$ be the matrix of latent variables in the model. Usually, owing to the complexity of the model, direct application of path sampling (Gelman and Meng, 1998) in evaluating the Bayes factor is difficult. In this chapter, we utilize the idea of data augmentation (Tanner and Wong, 1987) to solve the problem. Below we use similar reasonings as in Gelman and Meng (1998) to show briefly that path sampling can be applied to compute the Bayes factor by augmenting $\mathbf{Y}$ with $\boldsymbol{\Omega}$ and $\mathbf{L}$. From the equality $p(\boldsymbol{\Omega}, \mathbf{L}, \boldsymbol{\theta}|\mathbf{Y}) = p(\mathbf{Y}, \boldsymbol{\Omega}, \mathbf{L}, \boldsymbol{\theta})/p(\mathbf{Y})$, the marginal density $p(\mathbf{Y})$ can be treated as the normalizing constant of $p(\boldsymbol{\Omega}, \mathbf{L}, \boldsymbol{\theta}|\mathbf{Y})$, with the complete-data probability density $p(\mathbf{Y}, \boldsymbol{\Omega}, \mathbf{L}, \boldsymbol{\theta})$ taking as the unnormalized density. Now, consider the following class of densities which are denoted by a continuous parameter t in [0, 1]:

$$p(\boldsymbol{\Omega}, \mathbf{L}, \boldsymbol{\theta}|\mathbf{Y}, t) = \frac{1}{z(t)} p(\mathbf{Y}, \boldsymbol{\Omega}, \mathbf{L}, \boldsymbol{\theta}|t), \tag{8}$$

where

$$z(t) = p(\mathbf{Y}|t) = \int p(\mathbf{Y}, \boldsymbol{\Omega}, \mathbf{L}, \boldsymbol{\theta}|t)\, \mathrm{d}\boldsymbol{\Omega}\, \mathrm{d}\mathbf{L}\, \mathrm{d}\boldsymbol{\theta} = \int p(\mathbf{Y}, \boldsymbol{\Omega}, \mathbf{L}|\boldsymbol{\theta}, t) p(\boldsymbol{\theta})\, \mathrm{d}\boldsymbol{\Omega}\, \mathrm{d}\mathbf{L}\, \mathrm{d}\boldsymbol{\theta}, \tag{9}$$

with $p(\boldsymbol{\theta})$ being the prior density of $\boldsymbol{\theta}$, which is assumed to be independent of t.

In computing the Bayes factor, we construct a path using the parameter t in [0, 1] to link two competing models M_1 and M_0 together, so that $B_{10} = z(1)/z(0)$. Taking logarithm and then differentiating (9) with respect to t, and assuming the legitimacy of interchange of integration with differentiation, we have

$$\begin{aligned}
\frac{\mathrm{d} \log z(t)}{\mathrm{d}t} &= \int \frac{1}{z(t)} \frac{\mathrm{d}}{\mathrm{d}t} p(\mathbf{Y}, \boldsymbol{\Omega}, \mathbf{L}, \boldsymbol{\theta}|t)\, \mathrm{d}\boldsymbol{\Omega}\, \mathrm{d}\mathbf{L}\, \mathrm{d}\boldsymbol{\theta} \\
&= \int \frac{\mathrm{d}}{\mathrm{d}t} \log p(\mathbf{Y}, \boldsymbol{\Omega}, \mathbf{L}, \boldsymbol{\theta}|t) \cdot p(\boldsymbol{\Omega}, \mathbf{L}, \boldsymbol{\theta}|\mathbf{Y}, t)\, \mathrm{d}\boldsymbol{\Omega}\, \mathrm{d}\mathbf{L}\, \mathrm{d}\boldsymbol{\theta} \\
&= E_{\Omega, L, \theta} \left[\frac{\mathrm{d}}{\mathrm{d}t} \log p(\mathbf{Y}, \boldsymbol{\Omega}, \mathbf{L}, \boldsymbol{\theta}|t) \right],
\end{aligned}$$

where $E_{\Omega, L, \theta}$ denotes the expectation with respect to the distribution $p(\boldsymbol{\Omega}, \mathbf{L}, \boldsymbol{\theta}|\mathbf{Y}, t)$. Let

$$U(\mathbf{Y}, \boldsymbol{\Omega}, \mathbf{L}, \boldsymbol{\theta}, t) = \frac{\mathrm{d}}{\mathrm{d}t} \log p(\mathbf{Y}, \boldsymbol{\Omega}, \mathbf{L}, \boldsymbol{\theta}|t) = \frac{\mathrm{d}}{\mathrm{d}t} \log p(\mathbf{Y}, \boldsymbol{\Omega}, \mathbf{L}|\boldsymbol{\theta}, t), \tag{10}$$

which does not involve the prior density $p(\boldsymbol{\theta})$, we have

$$\log B_{10} = \log \frac{z(1)}{z(0)} = \int_0^1 E_{\Omega, L, \theta} [U(\mathbf{Y}, \boldsymbol{\Omega}, \mathbf{L}, \boldsymbol{\theta}, t)]\, \mathrm{d}t.$$

We follow the method as in Ogata (1989) to numerically evaluate the integral over t. Specifically, we first order the unique values of S fixed grids $\{t_{(s)}\}_{s=0}^{S}$ such that $t_{(0)} = 0 < t_{(1)} < t_{(2)} < \cdots < t_{(S)} < t_{(S+1)} = 1$, and estimate $\log B_{10}$ by

$$\widehat{\log B_{10}} = \frac{1}{2} \sum_{s=0}^{S} \left(t_{(s+1)} - t_{(s)}\right) \left(\bar{U}_{(s+1)} + \bar{U}_{(s)}\right), \tag{11}$$

where $\overline{U}_{(s)}$ is the average of the values of $U(\mathbf{Y}, \boldsymbol{\Omega}, \mathbf{L}, \boldsymbol{\theta}, t)$ for simulation draws at $t = t_{(s)}$, that is,

$$\bar{U}_{(s)} = J^{-1} \sum_{j=1}^{J} U\left(\mathbf{Y}, \boldsymbol{\Omega}^{(j)}, \mathbf{L}^{(j)}, \boldsymbol{\theta}^{(j)}, t_{(s)}\right), \tag{12}$$

in which $\left\{(\boldsymbol{\Omega}^{(j)}, \mathbf{L}^{(j)}, \boldsymbol{\theta}^{(j)}),\ j = 1, \ldots, J\right\}$ are simulated observations drawn from $p(\boldsymbol{\Omega}, \mathbf{L}, \boldsymbol{\theta} | \mathbf{Y}, t_{(s)})$.

In a comprehensive comparative study of various computing methods in computing the Bayes factor, Diciccio et al. (1997) pointed out that the bridge sampling (Meng and Wong, 1996) is an attractive method. Gelman and Meng (1998) showed that path sampling is a generalization of bridge sampling and importance sampling. Hence, it is expected that path sampling can give a more accurate result than bridge sampling in computing the Bayes factor. See Lee (2007) for applications of path sampling to various SEMs. In the next two sections, we illustrative path sampling through its applications to nonlinear SEMs and an integrated SEM.

4 Model Comparison of Nonlinear SEMs

4.1 Model Description

Consider a SEM with a random vector $\mathbf{y}_i (p \times 1)$ that satisfies the following measurement equation:

$$\mathbf{y}_i = \mathbf{A}\mathbf{c}_{yi} + \boldsymbol{\Lambda}\boldsymbol{\omega}_i + \boldsymbol{\epsilon}_i, \quad i = 1, \ldots, n, \tag{13}$$

where $\mathbf{A}(p \times c_1)$ and $\boldsymbol{\Lambda}(p \times q)$ are unknown parameter matrices, $\mathbf{c}_{yi} (\mathbf{c}_1 \times 1)$ is a vector of fixed covariates, $\boldsymbol{\omega}_i (q \times 1)$ is a vector of latent variables, $\boldsymbol{\epsilon}_i (p \times 1)$ is a vector of error measurements with distribution $N(\mathbf{0}, \boldsymbol{\Psi}_\epsilon)$, $\boldsymbol{\Psi}_\epsilon = \text{diag}(\psi_{\epsilon 1}, \ldots, \psi_{\epsilon p})$, $\boldsymbol{\omega}_i$ and $\boldsymbol{\epsilon}_i$ are independent. The latent vector $\boldsymbol{\omega}_i$ is partitioned into $(\boldsymbol{\eta}_i^T, \boldsymbol{\xi}_i^T)^T$ and is further modeled with an additional vector of fixed covariates $\mathbf{c}_i (\mathbf{c}_2 \times 1)$ through the following nonlinear structural equation:

$$\boldsymbol{\eta}_i = \mathbf{B}\mathbf{c}_i + \boldsymbol{\Pi}\boldsymbol{\eta}_i + \boldsymbol{\Gamma}\mathbf{F}(\boldsymbol{\xi}_i) + \boldsymbol{\delta}_i, \quad i = 1, \ldots, n, \tag{14}$$

where $\boldsymbol{\eta}_i(q_1 \times 1)$ and $\boldsymbol{\xi}_i(q_2 \times 1)$ are latent random vectors; $\mathbf{F}(\boldsymbol{\xi}_i) = (f_1(\boldsymbol{\xi}_i), \ldots, f_r(\boldsymbol{\xi}_i))^T$ is a $r \times 1$ vector-valued function with differentiable functions $f_1, \ldots, f_r$; $\mathbf{B}(q_1 \times c_2)$, $\boldsymbol{\Pi}(q_1 \times q_1)$, and $\boldsymbol{\Gamma}(q_1 \times r)$ are unknown parameter matrices. Moreover, it is assumed that $\boldsymbol{\Pi}_0 = \mathbf{I}_{q_1} - \boldsymbol{\Pi}$ is nonsingular; $|\boldsymbol{\Pi}_0|$ is independent of the elements in $\boldsymbol{\Pi}$; $\boldsymbol{\xi}$ and $\boldsymbol{\delta}$ are independently distributed as $N(\mathbf{0}, \boldsymbol{\Phi})$ and $N(\mathbf{0}, \boldsymbol{\Psi}_\delta)$, respectively; and $\boldsymbol{\Psi}_\delta = \mathrm{diag}(\psi_{\delta 1}, \ldots, \psi_{\delta q_1})$. In this model, nonlinear effects of the exogenous latent variables on the endogenous latent variables are included. The fixed covariates $\mathbf{c}_{yi}$ and $\mathbf{c}_i$ could be discrete, ordered categorical, or continuous measurements.

The above nonlinear SEM may be overparameterized for certain applications. For example, it does not allow intercepts to exist in both the measurement and structural equations defined by (13) and (14). Moreover, the choice of $\mathbf{F}(\boldsymbol{\xi}_i)$ is not arbitrary. For example, neither $f(\boldsymbol{\xi}) = 0$ nor $f_1(\boldsymbol{\xi}) = f_2(\boldsymbol{\xi})$ are allowed. Finally, the covariance structure is not identified. In this chapter, we follow the common practice in structural equation modeling by fixing appropriate elements in $\boldsymbol{\Lambda}$ at preassigned values. In most applications of SEMs, the positions and the preassigned values of the fixed elements in $\boldsymbol{\Lambda}$ are available from the subject knowledge or the purpose of the study. The parameter vector of the model includes all unknown structural parameters in $\mathbf{A}$, $\boldsymbol{\Lambda}$, $\boldsymbol{\Pi}$, $\boldsymbol{\Gamma}$, $\boldsymbol{\Psi}_\epsilon$, $\boldsymbol{\Psi}_\delta$, and $\boldsymbol{\Phi}$. In the following analysis, we assume that the nonlinear SEM under discussion is identified.

4.2 Model Comparison via Bayes Factor

Model comparison is an important issue in analyzing nonlinear SEMs. For instance, a fundamental problem is to decide whether a nonlinear structural equation is better than a linear structural equation in formulating a model to fit a given data set. To apply the path sampling procedure in computing the Bayes factor for model comparison, it is necessary to define a path t in [0, 1] to link the competing models. For most practical applications, a natural path is usually available.

To give a more specific illustration in applying the procedure to the current nonlinear SEM, consider the following models M_1 and M_0 that satisfy the same measurement equation (13), but with the following different nonlinear structural equations:

$$M_1 : \boldsymbol{\eta}_i = \mathbf{B}\mathbf{c}_i + \boldsymbol{\Pi}\boldsymbol{\eta}_i + \boldsymbol{\Gamma}_1\mathbf{F}_1(\boldsymbol{\xi}_i) + \boldsymbol{\delta}_i,$$
$$M_0 : \boldsymbol{\eta}_i = \mathbf{B}\mathbf{c}_i + \boldsymbol{\Pi}\boldsymbol{\eta}_i + \boldsymbol{\Gamma}_0\mathbf{F}_0(\boldsymbol{\xi}_i) + \boldsymbol{\delta}_i,$$

where $\mathbf{F}_1$ and $\mathbf{F}_0$ may involve different numbers of distinct nonlinear functions of $\boldsymbol{\xi}$; and $\boldsymbol{\Gamma}_1$ and $\boldsymbol{\Gamma}_0$ are the corresponding unknown coefficient matrices. In this illustration, differences of M_1 and M_0 are on the nonlinear relationships among the latent variables. The structural equations of model M_1 and M_0 are linked up by t in [0, 1] as follows:

$$M_t : \boldsymbol{\eta}_i = \mathbf{B}\mathbf{c}_i + \boldsymbol{\Pi}\boldsymbol{\eta}_i + (1-t)\boldsymbol{\Gamma}_0\mathbf{F}_0(\boldsymbol{\xi}_i) + t\boldsymbol{\Gamma}_1\mathbf{F}_1(\boldsymbol{\xi}_i) + \boldsymbol{\delta}_i = \boldsymbol{\Lambda}_{t\omega}\mathbf{G}(\mathbf{c}_i, \boldsymbol{\omega}_i) + \boldsymbol{\delta}_i,$$

where

$$\mathbf{\Lambda}_{tw} = (\mathbf{B}, \mathbf{\Pi}, (1-t)\mathbf{\Gamma}_0, t\mathbf{\Gamma}_1) \text{ and } \mathbf{G}(\mathbf{c}_i, \boldsymbol{\omega}_i) = (\mathbf{c}_i^T, \boldsymbol{\eta}_i^T, \mathbf{F}_0(\boldsymbol{\xi}_i)^T, \mathbf{F}_1(\boldsymbol{\xi}_i)^T)^T.$$

Clearly, when $t = 0$, $M_t = M_0$; when $t = 1$, $M_t = M_1$. Let $\mathbf{\Omega} = (\boldsymbol{\omega}_1, \ldots, \boldsymbol{\omega}_n)$, we have

$$\log p(\mathbf{Y}, \mathbf{\Omega}|\boldsymbol{\theta}, t) = -\frac{1}{2}\Bigg\{(p+q)n\log(2\pi) + n\log|\mathbf{\Psi}_\epsilon| + n\log|\mathbf{\Psi}_\delta| + n\log|\mathbf{\Phi}| + \sum_{i=1}^n \boldsymbol{\xi}_i^T\mathbf{\Phi}^{-1}\boldsymbol{\xi}_i + \sum_{i=1}^n(\mathbf{y}_i - \mathbf{A}\mathbf{c}_{yi} - \mathbf{\Lambda}\boldsymbol{\omega}_i)^T\,\mathbf{\Psi}_\epsilon^{-1} \times(\mathbf{y}_i - \mathbf{A}\mathbf{c}_{yi} - \mathbf{\Lambda}\boldsymbol{\omega}_i) + \sum_{i=1}^n(\boldsymbol{\eta}_i - \mathbf{\Lambda}_{t\omega}\mathbf{G}(\mathbf{c}_i, \boldsymbol{\omega}_i))^T\,\mathbf{\Psi}_\delta^{-1} \times(\boldsymbol{\eta}_i - \mathbf{\Lambda}_{t\omega}\mathbf{G}(\mathbf{c}_i, \boldsymbol{\omega}_i))\Bigg\}.$$

Here, $\boldsymbol{\theta}$ is the parameter vector in the linked model M_t. It contains all the common and distinct unknown parameters in M_0 and M_1; that is, unknown parameters in $\mathbf{A}$, $\mathbf{\Lambda}$, $\mathbf{\Psi}_\epsilon$, $\mathbf{B}$, $\mathbf{\Pi}$, $\mathbf{\Gamma}_0$, $\mathbf{\Gamma}_1$, $\mathbf{\Phi}$, and $\mathbf{\Psi}_\delta$. By differentiation with respect to t, we have

$$U(\mathbf{Y}, \mathbf{\Omega}, \boldsymbol{\theta}, t) = \frac{\mathrm{d}}{\mathrm{d}t}\log p(\mathbf{Y}, \mathbf{\Omega}|\boldsymbol{\theta}, t) = \sum_{i=1}^n(\boldsymbol{\eta}_i - \mathbf{\Lambda}_{t\omega}\mathbf{G}(\mathbf{c}_i, \boldsymbol{\omega}_i))^T\,\mathbf{\Psi}_\delta^{-1}\mathbf{\Lambda}_0\mathbf{G}(\mathbf{c}_i, \boldsymbol{\omega}_i), \tag{15}$$

where $\mathbf{\Lambda}_0 = (\mathbf{0}, \mathbf{0}, -\mathbf{\Gamma}_0, \mathbf{\Gamma}_1)$.

Application of this procedure to other special cases of the general model is similar. The main computation is in generating observations $(\mathbf{\Omega}^{(j)}, \boldsymbol{\theta}^{(j)})$ from $p(\mathbf{\Omega}, \boldsymbol{\theta}|\mathbf{Y}, t_{(s)})$ for evaluating $\bar{U}_{(s)}$. See Sect. 4.3 for an example with detailed implementation.

Observations $(\mathbf{\Omega}^{(j)}, \boldsymbol{\theta}^{(j)})$ from $p(\mathbf{\Omega}, \boldsymbol{\theta}|\mathbf{Y}, t_{(s)})$ are simulated by the Gibbs sampler (Geman and Geman, 1984) which is implemented as follows: At the jth iteration with a current $\boldsymbol{\theta}^{(j)}$ (a) generate $\mathbf{\Omega}^{(j+1)}$ from $p(\mathbf{\Omega}|\boldsymbol{\theta}^{(j)}, \mathbf{Y}, t)$, then (b) generate $\mathbf{\Omega}^{(j+1)}$ from $p(\boldsymbol{\theta}|\mathbf{\Omega}^{(j+1)}, \mathbf{Y}, t)$. The conditional distribution $p(\boldsymbol{\theta}|\mathbf{\Omega}, \mathbf{Y}, t)$ depends on the prior distribution of $\boldsymbol{\theta}$. Inspired by most Bayesian analyses of SEMs, the conjugate prior distributions are used. Let $\mathbf{\Lambda}_\omega = (\mathbf{B}, \mathbf{\Pi}, \mathbf{\Gamma})$, and let $\mathbf{A}_k^T$, $\mathbf{\Lambda}_k^T$, and $\mathbf{\Lambda}_{\omega k}^T$ be the kth rows of $\mathbf{A}$, $\mathbf{\Lambda}$, and $\mathbf{\Lambda}_\omega$, respectively. The conjugate prior distributions are given as follows:

$$p(\psi_{\epsilon k}^{-1}) \stackrel{D}{=} \text{Gamma}[\alpha_{0\epsilon k}, \beta_{0\epsilon k}],\ p(\mathbf{\Lambda}_k|\psi_{\epsilon k}) \stackrel{D}{=} N[\mathbf{\Lambda}_{0k}, \psi_{\epsilon k}\mathbf{H}_{0yk}],\ k = 1, \ldots, p,$$
$$p(\psi_{\delta k}^{-1}) \stackrel{D}{=} \text{Gamma}[\alpha_{0\delta k}, \beta_{0\delta k}],\ p(\mathbf{\Lambda}_{\omega k}|\psi_{\delta k}) \stackrel{D}{=} N[\mathbf{\Lambda}_{0\omega k}, \psi_{\delta k}\mathbf{H}_{0\omega k}],\ k = 1, \ldots, q,$$
$$p(\mathbf{\Phi}^{-1}) \stackrel{D}{=} W[\mathbf{R}_0, \rho_0],\ p(\mathbf{A}_k) \stackrel{D}{=} N[\mathbf{A}_{0k}, \mathbf{H}_{0k}],\ k = 1, \ldots, p, \tag{16}$$

where Gamma$[\alpha, \beta]$ represents the gamma distribution with a shape parameter $\alpha > 0$ and an inverse scale parameter $\beta > 0$; $\psi_{\epsilon k}$ and $\psi_{\delta k}$ are the kth diagonal elements of $\boldsymbol{\Psi}_\epsilon$ and $\boldsymbol{\Psi}_\delta$, respectively; $W[\cdot, \cdot]$ denotes the Wishart distribution; $\alpha_{0\epsilon k}, \beta_{0\epsilon k}, \alpha_{0\delta k}, \beta_{0\delta k}, \mathbf{A}_{0k}, \boldsymbol{\Lambda}_{0k}, \boldsymbol{\Lambda}_{0\omega k}, \rho_0$, and positive definite matrices $\mathbf{H}_{0k}$, $\mathbf{H}_{0yk}, \mathbf{H}_{0\omega k}$, and $\mathbf{R}_0$ are hyper-parameters, whose values are either subjectively determined if good prior information is available or objectively determined from the data or other sources. The full conditional distributions under these conjugate prior distributions can be obtained from Lee and Song (2003); or they can be obtained as special cases of the full conditional distributions presented in the Appendix.

4.3 A Simulation Study

The objectives of this simulation study are to reveal the performance of the path sampling procedure for computing Bayes factor and to evaluate the sensitivity with respect to prior inputs. Random observations were simulated from the nonlinear model defined by (13) and (14) with eight manifest variables, which are related to two fixed covariates $\{c_{yi1}, c_{yi2}\}$ and three latent variables $\{\eta_i, \xi_{i1}, \xi_{i2}\}$. The first fixed covariate c_{yi1} is sampled from a multinomial distribution which takes values 1.0, 2.0, and 3.0 with probabilities $\Phi^*(-0.5)$, $\Phi^*(0.5) - \Phi^*(-0.5)$, and $1.0 - \Phi^*(0.5)$, respectively, where Φ^* is the distribution function of $N[0, 1]$. The second covariate c_{yi2} is sampled from $N[0, 1]$. The true population values in matrices $\mathbf{A}$, $\boldsymbol{\Lambda}$, and $\boldsymbol{\Psi}_\epsilon$ are

$$\mathbf{A}^T = \begin{bmatrix} 1.0\ 1.0\ 1.0\ 1.0\ 1.0\ 1.0\ 1.0\ 1.0 \\ 0.7\ 0.7\ 0.7\ 0.7\ 0.7\ 0.7\ 0.7\ 0.7 \end{bmatrix},$$

$$\boldsymbol{\Lambda}^T = \begin{bmatrix} 1.0^* & 1.5 & 1.5 & 0.0^* & 0.0^* & 0.0^* & 0.0^* & 0.0^* \\ 0.0^* & 0.0^* & 0.0^* & 1.0^* & 1.5 & 0.0^* & 0.0^* & 0.0^* \\ 0.0^* & 0.0^* & 0.0^* & 0.0^* & 0.0^* & 1.0^* & 1.5 & 1.5 \end{bmatrix}, \quad \boldsymbol{\Psi}_\varepsilon = \mathbf{I}_8,$$

where $\mathbf{I}_8$ is a 8 by 8 identity matrix, and parameters with an asterisk were treated as known. The true variances and covariance of ξ_{i1} and ξ_{i2} are $\phi_{11} = \phi_{22} = 1.0$, and $\phi_{21} = 0.15$. These two latent variables are related to η_i by

$$\eta_i = 1.0c_i + 0.5\xi_{i1} + 0.5\xi_{i2} + 1.0\xi_{i2}^2 + \delta_i,$$

where c_i is another fixed covariate sampled from a Bernoulli distribution that takes 1.0 with probability 0.7 and 0.0 with probability 0.3; and $\psi_\delta = 1.0$. On the basis of these specifications, random samples $\{\mathbf{y}_i, i = 1, \ldots, n\}$ with $n = 300$ were generated for the simulation study. A total of 100 replications were taken for each case.

Attention is devoted to compare models with different formations of the more interesting structural equation with latent variables. Hence, models with the same measurement equation and the following structural equations are involved in the model comparison:

$$M_0: \ \eta_i = bc_i + \gamma_1\xi_{i1} + \gamma_2\xi_{i2} + \gamma_{22}\xi_{i2}^2 + \delta_i,$$
$$M_1: \ \eta_i = bc_i + \gamma_1\xi_{i1} + \gamma_2\xi_{i2} + \delta_i,$$
$$M_2: \ \eta_i = bc_i + \gamma_1\xi_{i1} + \gamma_2\xi_{i2} + \gamma_{12}\xi_{i1}\xi_{i2} + \delta_i,$$
$$M_3: \ \eta_i = bc_i + \gamma_1\xi_{i1} + \gamma_2\xi_{i2} + \gamma_{11}\xi_{i1}^2 + \delta_i,$$
$$M_4: \ \eta_i = bc_i + \gamma_1\xi_{i1} + \gamma_2\xi_{i2} + \gamma_{12}\xi_{i1}\xi_{i2} + \gamma_{11}\xi_{i1}^2 + \delta_i,$$
$$M_5: \ \eta_i = \gamma_1\xi_{i1} + \gamma_2\xi_{i2} + \gamma_{22}\xi_{i2}^2 + \delta_i,$$
$$M_6: \ \eta_i = bc_i + \gamma_1\xi_{i1} + \gamma_2\xi_{i2} + \gamma_{12}\xi_{i1}\xi_{i2} + \gamma_{11}\xi_{i1}^2 + \gamma_{22}\xi_{i2}^2 + \delta_i.$$

Here, M_0 is the true model; M_1 is a linear model; M_2, M_3, and M_4 are nonnested in M_0; and M_0 is nested in the most general model M_6. To give a more detailed illustration in applying the procedure to model comparison of nonlinear SEMs, the implementation of path sampling to estimate $\log B_{02}$ in comparing M_0 and M_2 is given here. Let $\boldsymbol{\theta} = (\tilde{\boldsymbol{\theta}}, \boldsymbol{\Lambda}_\omega)$ and $\boldsymbol{\theta}_t = (\tilde{\boldsymbol{\theta}}, \boldsymbol{\Lambda}_{t\omega})$, where $\boldsymbol{\Lambda}_\omega = (b, \gamma_1, \gamma_2, \gamma_{12}, \gamma_{22})$, $\boldsymbol{\Lambda}_{t\omega} = (b, \gamma_1, \gamma_2, (1-t)\gamma_{12}, t\gamma_{22})$, and $\tilde{\boldsymbol{\theta}}$ includes all unknown parameters in M_t except for $\boldsymbol{\Lambda}_\omega$. The procedure consists of the following steps:

Step 1: Select a M_t to link M_0 and M_2. Here, M_t is defined with the same measurement model as in M_0 and M_2, but with the following structural equation:

$$M_t: \eta_i = bc_i + \gamma_1\xi_{i1} + \gamma_2\xi_{i2} + (1-t)\gamma_{12}\xi_{i1}\xi_{i2} + t\gamma_{22}\xi_{i2}^2 + \delta_i.$$

Clearly, when $t = 1$, $M_t = M_0$; when $t = 0$, $M_t = M_2$.

Step 2: At the fixed grid $t = t_{(s)}$, generate observations $(\boldsymbol{\Omega}^{(j)}, \boldsymbol{\theta}^{(j)})$, $j = 1, \ldots, J$ from $p(\boldsymbol{\Omega}, \boldsymbol{\theta}|\mathbf{Y}, t_{(s)})$. Specifically, at the jth iteration

1. Generate $\boldsymbol{\Omega}^{(j+1)}$ from $p(\boldsymbol{\Omega}|\boldsymbol{\theta}_t^{(j)}, \mathbf{Y})$.
2. Generate $\tilde{\boldsymbol{\theta}}^{(j+1)}$ from $p(\tilde{\boldsymbol{\theta}}|\boldsymbol{\Omega}^{(j+1)}, \boldsymbol{\Lambda}_{t\omega}^{(j)}, \mathbf{Y})$.
3. Generate $\boldsymbol{\Lambda}_\omega^{(j+1)}$ from $p(\boldsymbol{\Lambda}_\omega|\boldsymbol{\Omega}^{(j+1)}, \tilde{\boldsymbol{\theta}}^{(j+1)}, \mathbf{Y})$.
4. Update $\boldsymbol{\theta}_t^{(j+1)}$ by letting $\boldsymbol{\theta}_t^{(j+1)} = (\tilde{\boldsymbol{\theta}}^{(j+1)}, \boldsymbol{\Lambda}_{t\omega}^{(j+1)})$, where $\boldsymbol{\Lambda}_{t\omega}^{(j+1)} = (1, 1, 1, (1-t_{(s)}), t_{(s)})\boldsymbol{\Lambda}_\omega^{(j+1)^T}$.

Step 3: Calculate $U(\mathbf{Y}, \boldsymbol{\Omega}^{(j)}, \boldsymbol{\theta}^{(j)}, t_{(s)})$ by substituting $\{(\boldsymbol{\Omega}^{(j)}, \boldsymbol{\theta}^{(j)}); j = 1, \ldots, J\}$ to (15) as follows:

$$U(\mathbf{Y}, \boldsymbol{\Omega}, \boldsymbol{\theta}, t_{(s)}) = \sum_{i=1}^{n} (\eta_i - bc_i - \gamma_1\xi_1 - \gamma_2\xi_2 - (1-t_{(s)})\gamma_{12}\xi_1\xi_2 - t_{(s)}\gamma_{22}\xi_2^2)$$
$$\times(\gamma_{12}\xi_1\xi_2 - \gamma_{22}\xi_2^2)/\psi_\delta.$$

Step 4: Calculate $\bar{U}_{(s)}$; see (12).

Step 5: Repeat Step 2 to Step 5 until all $\bar{U}_{(s)}, s = 0, \ldots, S$ are calculated. Then, $\widehat{\log B_{02}}$ can be estimated by (11).

In the sensitivity analysis concerning about the prior inputs, the less important hyper-parameters in the conjugate prior distribution are selected as $\mathbf{H}_{0k} = \mathbf{I}$,

$\mathbf{H}_{0yk} = \mathbf{I}$, and $\mathbf{H}_{0\omega k} = \mathbf{I}$. For the more important hyper-parameters, we followed the suggestion of Kass and Raftery (1995) to perturb them as follows. For $\alpha_{0\epsilon k} = \alpha_{0\delta k} = 8$, $\beta_{0\epsilon k} = \beta_{0\delta k} = 10$, and $\rho_0 = 20$, we consider the following three types of prior inputs:

(I) $\mathbf{A}_{0k}$, $\mathbf{\Lambda}_{0k}$, and $\mathbf{\Lambda}_{0\omega k}$ are selected to be the true parameter matrices, and $\mathbf{R}_0^{-1} = (\rho_0 - q_2 - 1)\mathbf{\Phi}_0$, where elements in $\mathbf{\Phi}_0$ are the true parameter values.
(II) The hyper-parameters specified in (I) are equal to half of the values given in (I).
(III) The hyper-parameters specified in (I) are equal to twice of the values given in (I).
For Type (I) prior inputs as given above, we consider the following prior inputs on $\alpha_{0\epsilon k}$, $\alpha_{0\delta k}$, $\beta_{0\epsilon k}$, $\beta_{0\delta k}$, and ρ_0:
(IV) $\alpha_{0\epsilon k} = \alpha_{0\delta k} = 3$, $\beta_{0\epsilon k} = \beta_{0\delta k} = 5$, and $\rho_0 = 12$.
(V) $\alpha_{0\epsilon k} = \alpha_{0\delta k} = 12$, $\beta_{0\epsilon k} = \beta_{0\delta k} = 15$, and $\rho_0 = 30$.

For every case, we took 20 grids in [0, 1] and collected $J = 1,000$ iterations after discarding 500 burn-in iterations at each grid in the computation of Bayes factor. σ^2 was set to be 1.0 in the MH algorithm, which gives an approximate acceptance rate 0.43. Estimates of $\log B_{0k}$, $k = 1, \ldots, 6$ under the three different priors were computed. The mean and standard deviation of $\widehat{\log B_{0k}}$ were also computed on the basis of 100 replications. Results corresponding to $\widehat{\log B_{0k}}$, $k = 1, \ldots, 5$ and $\widehat{B_{60}}$ are reported in Table 1. Moreover, for each $k = 1, \ldots, 6$, we evaluate

$$D(\text{I} - \text{II}) = \max\left\{|\widehat{\log B_{0k}}(\text{I}) - \widehat{\log B_{0k}}(\text{II})|\right\}$$

as well as D(I–III) and D(IV–V) similarly, where $\widehat{\log B_{0k}}(\text{I})$ is the estimate under prior (I) and so on, and "max" is the maximum taken over the 100 replications. The results are presented in Table 2, for example, the maximum difference of the estimates of $\log B_{01}$ obtained via priors (I) and (II) is 6.55. From the rows of Table 1, we observe that the means and standard deviations of $\widehat{\log B_{0k}}$ obtained under different prior inputs are close to each other. This indicates that the estimate of $\log B_{0k}$ is not very sensitive to these prior inputs under a sample size of 300. For practical applications, we see from Table 2 that even for the worst situation with the

Table 1 Mean and standard errors of the estimated $\log B_{0k}$ in the simulation study

	Mean (std)				
	Prior I	Prior II	Prior III	Prior IV	Prior V
$\log B_{01}$	106.28 (25.06)	107.58 (25.15)	102.96 (24.81)	103.87 (22.71)	104.61 (23.92)
$\log B_{02}$	102.16 (24.91)	103.45 (25.02)	99.17 (24.54)	99.98 (22.67)	100.49 (23.47)
$\log B_{03}$	109.51 (25.63)	111.23 (25.74)	105.96 (25.19)	107.20 (23.81)	108.24 (24.59)
$\log B_{04}$	105.23 (25.31)	106.61 (25.47)	101.83 (24.90)	103.16 (23.78)	103.69 (24.12)
$\log B_{05}$	17.50 (5.44)	18.02 (5.56)	16.65 (5.21)	18.02 (5.34)	17.85 (5.30)
$\log B_{60}$	0.71 (0.54)	0.71 (0.51)	0.69 (0.55)	0.78 (0.67)	0.75 (0.65)

Table 2 Maximum absolute differences of $\log B_{0k}$ under some different priors

	$\log B_{01}$	$\log B_{02}$	$\log B_{03}$	$\log B_{04}$	$\log B_{05}$	$\log B_{60}$
D(I–II)	6.55	5.47	8.22	5.24	2.18	0.27
D(I–III)	7.84	9.33	10.23	10.17	3.07	0.31
D(IV–V)	14.03	17.86	13.65	4.87	1.91	0.25

maximum absolute deviation, the estimated logarithm of Bayes factors under different prior inputs give the same conclusion for selecting the model via the criterion given in (1).

From Table 1, it is clear that M_0 is much better than the linear model M_1 and the nonnested models M_2, M_3, M_4, and M_5. Thus, the correct model is selected. For comparison with the encompassing model M_6, we found that out of 100 replications under prior (I), 75 of the $\widehat{\log B_{60}}$ are in the interval (0.0, 1.0), 23 of them are in (1.0, 2.0), and only 2 of them are in (2.0, 3.0). Since M_0 is simpler than M_6, it should be selected if $\widehat{\log B_{60}}$ is in (0.0, 1.0). Thus, the true model is selected in 75 out of the 100 replications. Owing to randomness, the remaining $\widehat{\log B_{60}}$ support mildly the encompassing model. Although the encompassing model is not the true model, it should not be regarded as an incorrect model for fitting the data. It has been pointed out by Kass and Raftery (1995) that the effect of the priors is small in estimation. To give some ideas about the empirical performance of the proposed procedure on estimation, the means of the Bayesian estimates and the root mean squares (RMS) between the Bayesian estimates and the true values of M_0 over the 100 replications under some prior inputs are reported in Table 3. It seems that Bayesian estimates are quite accurate and not very sensitive to the selected prior inputs.

5 Model Comparison of an Integrated SEM

Motivated by the demand of efficient statistical methods for analyzing various kinds of complex real data in substantive research, the recent growth of SEM has been rather repaid. Efficient methods have been developed to handle missing data (see Dolan et al., 2005; Song and Lee, 2006b), dichotomous or ordered categorical data (Song and Lee, 2004, 2005), and hierarchical or multilevel data (Ansari and Jedidi, 2000; Raykov and Marcoulides, 2006; Lee and Song, 2004b). Although model comparison has been separately addressed in some of the above mentioned articles, however, it is necessary to compute the Bayes factor in the context of an integrated model for model comparison under some complex situations. To see this point, let us consider, for example, the problem of comparing a nonlinear SEM (M_1) with a two-level linear SEM (M_2). The computational method that was developed based on M_1 for computing the Bayes factor can only be used to compare models under the model framework of the nonlinear SEM model. As the method developed under

Table 3 Mean and RMS of Bayesian estimates under M_0 with different priors

Para	Prior I		Prior II		Prior III	
	Mean	RMS	Mean	RMS	Mean	RMS
$\lambda_{21} = 1.5$	1.495	0.047	1.491	0.048	1.498	0.047
$\lambda_{31} = 1.5$	1.493	0.049	1.490	0.049	1.497	0.049
$\lambda_{52} = 1.5$	1.467	0.127	1.467	0.120	1.465	0.119
$\lambda_{73} = 1.5$	1.525	0.094	1.525	0.098	1.536	0.104
$\lambda_{83} = 1.5$	1.534	0.098	1.533	0.101	1.544	0.104
$b = 1.0$	1.010	0.139	1.017	0.133	1.004	0.131
$\gamma_1 = 0.5$	0.518	0.091	0.519	0.092	0.516	0.091
$\gamma_2 = 0.5$	0.505	0.123	0.494	0.126	0.535	0.133
$\gamma_{22} = 1.0$	1.060	0.141	1.064	0.148	1.065	0.146
$a_{11} = 1.0$	1.000	0.064	0.992	0.062	1.015	0.065
$a_{21} = 1.0$	1.007	0.094	0.997	0.087	1.026	0.095
$a_{31} = 1.0$	1.000	0.093	0.990	0.086	1.019	0.091
$a_{41} = 1.0$	0.991	0.037	0.989	0.037	0.996	0.036
$a_{51} = 1.0$	0.992	0.046	0.989	0.047	0.999	0.045
$a_{61} = 1.0$	1.000	0.031	0.997	0.031	1.006	0.032
$a_{71} = 1.0$	0.997	0.045	0.993	0.046	1.006	0.047
$a_{81} = 1.0$	0.998	0.040	0.994	0.041	1.007	0.041
$a_{12} = 0.7$	0.701	0.095	0.685	0.095	0.726	0.096
$a_{22} = 0.7$	0.700	0.123	0.676	0.124	0.735	0.125
$a_{32} = 0.7$	0.703	0.108	0.680	0.110	0.739	0.113
$a_{42} = 0.7$	0.704	0.086	0.695	0.085	0.718	0.088
$a_{52} = 0.7$	0.723	0.102	0.710	0.098	0.741	0.107
$a_{62} = 0.7$	0.702	0.054	0.698	0.055	0.711	0.056
$a_{72} = 0.7$	0.695	0.068	0.690	0.069	0.707	0.068
$a_{82} = 0.7$	0.713	0.080	0.708	0.080	0.725	0.083
$\psi_{\epsilon 1} = 1.0$	0.842	0.095	0.841	0.094	0.840	0.094
$\psi_{\epsilon 2} = 1.0$	0.839	0.098	0.843	0.100	0.852	0.104
$\psi_{\epsilon 3} = 1.0$	0.836	0.093	0.839	0.094	0.847	0.096
$\psi_{\epsilon 4} = 1.0$	0.811	0.077	0.813	0.077	0.812	0.079
$\psi_{\epsilon 5} = 1.0$	0.948	0.179	0.945	0.176	0.949	0.179
$\psi_{\epsilon 6} = 1.0$	0.839	0.082	0.839	0.082	0.841	0.083
$\psi_{\epsilon 7} = 1.0$	0.852	0.094	0.851	0.094	0.851	0.094
$\psi_{\epsilon 8} = 1.0$	0.846	0.086	0.846	0.087	0.846	0.087
$\psi_{\delta} = 1.0$	1.021	0.109	1.028	0.111	1.026	0.112
$\phi_{11} = 1.0$	0.980	0.140	0.980	0.137	0.981	0.136
$\phi_{12} = .15$	0.148	0.076	0.148	0.076	0.147	0.076
$\phi_{22} = 1.0$	0.959	0.126	0.962	0.129	0.948	0.132

M_1 cannot be applied to a different two-level SEM, the model comparison problem cannot be solved. Similarly, even simultaneously given a separate Bayesian development on the basis of M_2, the computational procedure that is developed under a two-level linear SEM cannot handle a nonlinear SEM model. A solution to this problem requires the development of an integrated model that subsumes both M_1 and M_2, so that model comparison can be done under a comprehensive framework.

5.1 The Integrated Model

We consider a two-level nonlinear SEM with missing dichotomous and ordered categorical variable. In generic sense, the specific relationship of an ordered categorical variable z and its underlying continuous variable y is given by

$$z = k \quad \text{if} \quad \alpha_{k-1} \leq y < \alpha_k, \quad \text{for } k = 1, \ldots, m, \tag{17}$$

where $\{-\infty = \alpha_0 < \alpha_1 < \cdots < \alpha_{m-1} < \alpha_m = \infty\}$ is the set of unknown thresholds that defines m categories. The variance and thresholds corresponding to each ordered categorical variable are not identifiable. A common method for achieving identification is to fix the smallest and the largest thresholds, α_1 and α_{m-1}, of the corresponding ordered categorical variable at preassigned values (see, Shi and Lee, 2000), for example, $\alpha_1 = \Phi^{*-1}(f_1^*)$ and $\alpha_{m-1} = \Phi^{*-1}(f_{m-1}^*)$, where f_k^* is the observed cumulative marginal proportion of the categories with $z < k$. Like an ordered categorical variable, the link between a dichotomous variable, say d, with its underlying continuous variable, say w, can be represented by a threshold specification as follows:

$$d = 1 \quad \text{if } w > \alpha, \quad \text{and} \quad d = 0 \quad \text{if } w \leq \alpha, \tag{18}$$

where α is an unknown threshold parameter. To identify a dichotomous variable, we use the method suggested by Song and Lee (2005) by fixing the corresponding error measurement's variance at a preassigned value, for example, 1.0.

To formulate the integrated model, we consider a collection of p-variate random vectors $\mathbf{u}_{gi}$ for $i = 1, \ldots, N_g$, within groups $g = 1, \ldots, G$. As the sample sizes N_g may differ from group to group, the data set is unbalanced. We assume that conditional on the group mean $\mathbf{v}_g$, random observations in each group at the within-groups (first) level have the following structure:

$$\mathbf{u}_{gi} = \mathbf{v}_g + \mathbf{A}_{1g}\mathbf{c}_{ugi} + \boldsymbol{\Lambda}_{1g}\boldsymbol{\omega}_{1gi} + \boldsymbol{\epsilon}_{1gi}, \quad g = 1, \ldots, G, \ i = 1, \ldots, N_g, \tag{19}$$

where $\mathbf{c}_{ugi}$ is a vector of fixed covariates, $\mathbf{A}_{1g}$ is a matrix of coefficients, $\boldsymbol{\Lambda}_{1g}$ is a matrix of factor loadings, $\boldsymbol{\omega}_{1gi}$ is a $q_1 \times 1$ vector of latent variables, and $\boldsymbol{\epsilon}_{1gi}$ is a $p \times 1$ vector of error measurements. It is assumed that $\boldsymbol{\epsilon}_{1gi}$ is independent of $\boldsymbol{\omega}_{1gi}$ and is distributed as $N[\mathbf{0}, \boldsymbol{\Psi}_{1g}]$, where $\boldsymbol{\Psi}_{1g}$ is a diagonal matrix. At the between-groups (second) level, we assume that $\mathbf{v}_g$ has the structure

$$\mathbf{v}_g = \mathbf{A}_2\mathbf{c}_{vg} + \boldsymbol{\Lambda}_2\boldsymbol{\omega}_{2g} + \boldsymbol{\epsilon}_{2g}, \quad g = 1, \ldots, G, \tag{20}$$

where $\mathbf{c}_{vg}$ is a vector of fixed covariates, $\mathbf{A}_2$ is a matrix of coefficients, $\boldsymbol{\Lambda}_2$ is a matrix of factor loadings, $\boldsymbol{\omega}_{2g}$ is a $q_2 \times 1$ vector of latent variables, and $\boldsymbol{\epsilon}_{2g}$ is a $p \times 1$ vector of error measurements. It is assumed that $\boldsymbol{\epsilon}_{2g}$ is independent of $\boldsymbol{\omega}_{2g}$ and is distributed as $N[\mathbf{0}, \boldsymbol{\Psi}_2]$, where $\boldsymbol{\Psi}_2$ is a diagonal matrix. Moreover, the first level latent vectors are assumed to be independent of the second level latent vectors. However, because of the presence of $\mathbf{v}_g$, $\mathbf{u}_{gi}$ and $\mathbf{u}_{gj}$ are correlated, and the usual assumption about independence is violated. Equations (19) and (20) define the measurement equations for the within-groups and between-groups models. To assess the relationships among latent variables at both levels, the following nonlinear structural equations in the between-groups and within-groups models are considered:

$$\boldsymbol{\eta}_{1gi} = \mathbf{B}_{1g}\mathbf{c}_{1gi} + \boldsymbol{\Pi}_{1g}\boldsymbol{\eta}_{1gi} + \boldsymbol{\Gamma}_{1g}\mathbf{F}_1(\boldsymbol{\xi}_{1gi}) + \boldsymbol{\delta}_{1gi}, \text{ and} \tag{21}$$

$$\boldsymbol{\eta}_{2g} = \mathbf{B}_2\mathbf{c}_{2g} + \boldsymbol{\Pi}_2\boldsymbol{\eta}_{2g} + \boldsymbol{\Gamma}_2\mathbf{F}_2(\boldsymbol{\xi}_{2g}) + \boldsymbol{\delta}_{2g}, \tag{22}$$

where $\mathbf{c}_{1gi}$ and $\mathbf{c}_{2g}$ are fixed covariates, $\mathbf{B}_{1g}$, $\mathbf{B}_2$, $\boldsymbol{\Pi}_{1g}$, $\boldsymbol{\Pi}_2$, $\boldsymbol{\Gamma}_{1g}$, and $\boldsymbol{\Gamma}_2$ are matrices of coefficients, $\mathbf{F}_1(\boldsymbol{\xi}_{1gi}) = \left(f_{11}(\boldsymbol{\xi}_{1gi}), \ldots, f_{1a}(\boldsymbol{\xi}_{1gi})\right)^T$ and $\mathbf{F}_2(\boldsymbol{\xi}_{2g}) = \left(f_{21}(\boldsymbol{\xi}_{2g}), \ldots, f_{2b}(\boldsymbol{\xi}_{2g})\right)^T$ are vector-valued functions with nonzero differentiable functions. We assume that $\mathbf{I}_1 - \boldsymbol{\Pi}_{1g}$ and $\mathbf{I}_2 - \boldsymbol{\Pi}_2$ are nonsingular and their determinants are independent of $\boldsymbol{\Pi}_{1g}$ and $\boldsymbol{\Pi}_2$, respectively; $\boldsymbol{\xi}_{1gi}$ and $\boldsymbol{\delta}_{1gi}$ are independently distributed as $N[\mathbf{0}, \boldsymbol{\Phi}_{1g}]$ and $N[\mathbf{0}, \boldsymbol{\Psi}_{1\delta g}]$, respectively, where $\boldsymbol{\Psi}_{1\delta g}$ is a diagonal matrix. Similarly, it is assumed that $\boldsymbol{\xi}_{2g}$ and $\boldsymbol{\delta}_{2g}$ are independently distributed as $N[\mathbf{0}, \boldsymbol{\Phi}_2]$ and $N[\mathbf{0}, \boldsymbol{\Psi}_{2\delta}]$, respectively, where $\boldsymbol{\Psi}_{2\delta}$ is a diagonal matrix. However, owing to the nonlinear functions in $\mathbf{F}_1$ and $\mathbf{F}_2$, the distribution of $\mathbf{u}_{gi}$ is not normal.

To investigate the model with mixed continuous, dichotomous, and ordered categorical variables, we suppose without loss of generality that $\mathbf{u}_{gi} = (\mathbf{x}_{gi}^T, \mathbf{w}_{gi}^T, \mathbf{y}_{gi}^T)^T$, where $\mathbf{x}_{gi} = (x_{gi1}, \ldots, x_{gir})^T$ is an observable continuous random vector, $\mathbf{w}_{gi} = (w_{gi1}, \ldots, w_{gis})^T$ and $\mathbf{y}_{gi} = (y_{gi1}, \ldots, y_{git})^T$ are unobservable continuous random vectors that underlie the observable dichotomous and ordered categorical vectors $\mathbf{d}_{gi}$ and $\mathbf{z}_{gi}$, respectively. The links between an ordered categorical variable and a dichotomous variable with their underlying continuous variables are given by (17) and (18), respectively. Entries of $\mathbf{u}_{gi}$ are allowed to be missing at random. We identify the covariance models by fixing appropriate elements of $\boldsymbol{\Lambda}_{1g}$, $\boldsymbol{\Lambda}_2$, $\boldsymbol{\Pi}_{1g}$, $\boldsymbol{\Gamma}_{1g}$, $\boldsymbol{\Pi}_2$, and $\boldsymbol{\Gamma}_2$ at preassigned values.

5.2 Model Comparison

We first consider the posterior simulation to generate the required observations from the joint posterior distribution for computing the Bayes factor, see (11) and (12).

Let $\mathbf{X}_{\text{obs}}$, $\mathbf{D}_{\text{obs}}$, and $\mathbf{Z}_{\text{obs}}$ be the observed data corresponding to the continuous, dichotomous, and ordered categorical variables, and let $\mathbf{X}_{\text{mis}}$, $\mathbf{D}_{\text{mis}}$, and $\mathbf{Z}_{\text{mis}}$ be the missing data corresponding to these types of variables. It is assumed that missing data are missing at random. Let $\mathbf{W}_{\text{obs}}$ and $\mathbf{W}_{\text{mis}}$ be the underlying unobservable continuous measurements corresponding to $\mathbf{D}_{\text{obs}}$ and $\mathbf{D}_{\text{mis}}$; and let $\mathbf{Y}_{\text{obs}}$ and $\mathbf{Y}_{\text{mis}}$ be the underlying unobservable continuous measurements corresponding to $\mathbf{Z}_{\text{obs}}$ and $\mathbf{Z}_{\text{mis}}$, respectively. Let $\mathbf{O} = (\mathbf{X}_{\text{obs}}, \mathbf{D}_{\text{obs}}, \mathbf{Z}_{\text{obs}})$ be the observed data set; Moreover, let $\boldsymbol{\Omega}_{1g} = (\boldsymbol{\omega}_{1gi}, \ldots, \boldsymbol{\omega}_{1gN_g})$ and $\boldsymbol{\Omega}_1 = (\boldsymbol{\Omega}_{11}, \ldots, \boldsymbol{\Omega}_{1G})$ be the matrices that contain the within-groups latent vectors and matrices; and let $\boldsymbol{\Omega}_2 = (\boldsymbol{\omega}_{21}, \ldots, \boldsymbol{\omega}_{2G})$ and $\mathbf{V} = (\mathbf{v}_1, \ldots, \mathbf{v}_G)$ be the matrices that contain the between-groups latent vectors. Finally, let $\boldsymbol{\alpha}$ and $\boldsymbol{\theta}$ be the vectors that, respectively, contain all unknown thresholds and all unknown parameters that are involved in the model defined by (19)–(22). A sufficiently large number of observations will be simulated from the following joint posterior distribution $p(\boldsymbol{\theta}, \boldsymbol{\alpha}, \mathbf{Y}_{\text{obs}}, \mathbf{W}_{\text{obs}}, \mathbf{U}_{\text{mis}}, \boldsymbol{\Omega}_1, \boldsymbol{\Omega}_2, \mathbf{V}|\mathbf{O})$, where $\mathbf{U}_{\text{mis}} = (\mathbf{X}_{\text{mis}}, \mathbf{Y}_{\text{mis}}, \mathbf{Z}_{\text{mis}})$. The simulation is done by the Gibbs sampler, which iteratively draws samples from the following full conditional distributions: $p(\boldsymbol{\theta}|\boldsymbol{\alpha}, \mathbf{Y}_{\text{obs}}, \mathbf{W}_{\text{obs}}, \mathbf{U}_{\text{mis}}, \boldsymbol{\Omega}_1, \boldsymbol{\Omega}_2, \mathbf{V}, \mathbf{O})$, $p(\boldsymbol{\alpha}, \mathbf{Y}_{\text{obs}}|\boldsymbol{\theta}, \mathbf{W}_{\text{obs}}, \mathbf{U}_{\text{mis}}, \boldsymbol{\Omega}_1, \boldsymbol{\Omega}_2, \mathbf{V}, \mathbf{O})$, $p(\mathbf{W}_{\text{obs}}|\boldsymbol{\theta}, \boldsymbol{\alpha}, \mathbf{Y}_{\text{obs}}, \mathbf{U}_{\text{mis}}, \boldsymbol{\Omega}_1, \boldsymbol{\Omega}_2, \mathbf{V}, \mathbf{O})$, $p(\mathbf{U}_{\text{mis}}|\boldsymbol{\theta}, \boldsymbol{\alpha}, \mathbf{Y}_{\text{obs}}, \mathbf{W}_{\text{obs}}, \boldsymbol{\Omega}_1, \boldsymbol{\Omega}_2, \mathbf{V}, \mathbf{O})$, $p(\boldsymbol{\Omega}_1|\boldsymbol{\theta}, \boldsymbol{\alpha}, \mathbf{Y}_{\text{obs}}, \mathbf{W}_{\text{obs}}, \mathbf{U}_{\text{mis}}, \boldsymbol{\Omega}_2, \mathbf{V}, \mathbf{O})$, $p(\boldsymbol{\Omega}_2|\boldsymbol{\theta}, \boldsymbol{\alpha}, \mathbf{Y}_{\text{obs}}, \mathbf{W}_{\text{obs}}, \mathbf{U}_{\text{mis}}, \boldsymbol{\Omega}_1, \mathbf{V}, \mathbf{O})$, and $p(\mathbf{V}|\boldsymbol{\theta}, \boldsymbol{\alpha}, \mathbf{Y}_{\text{obs}}, \mathbf{W}_{\text{obs}}, \mathbf{U}_{\text{mis}}, \boldsymbol{\Omega}_1, \boldsymbol{\Omega}_2, \mathbf{O})$. Note that $p(\boldsymbol{\theta}|\boldsymbol{\alpha}, \mathbf{Y}_{\text{obs}}, \mathbf{W}_{\text{obs}}, \mathbf{U}_{\text{mis}}, \boldsymbol{\Omega}_1, \boldsymbol{\Omega}_2, \mathbf{V}, \mathbf{O}) = p(\boldsymbol{\theta}|*)$ is further decomposed into the following components: $p(\mathbf{A}_1|\boldsymbol{\theta}_{-A_1}, *)$, $p(\boldsymbol{\Lambda}_1|\boldsymbol{\theta}_{-\Lambda_1}, *), \ldots, p(\boldsymbol{\Psi}_{2\delta}|\boldsymbol{\theta}_{-\Psi_{2\delta}}, *)$, where $\boldsymbol{\theta}_{-A_1}, \boldsymbol{\theta}_{-\Lambda_1}, \ldots, \boldsymbol{\theta}_{-\Psi_{2\delta}}$ are subvectors of $\boldsymbol{\theta}$ without $\mathbf{A}_1, \boldsymbol{\Lambda}_1, \ldots, \boldsymbol{\Psi}_{2\delta}$, respectively. The above full conditional distributions required for implementing the Gibbs sampler are given in the Appendix. Some of the conditional distributions are standard distributions such as normal, univariate truncated normal, Gamma, and inverted Wishart, simulating observations from them is straight-forward and fast. The Metropolis–Hastings (MH) (Metropolis et al., 1953; Hastings, 1970) algorithm will be used to simulate observations from the following more complicated conditional distributions, $p(\boldsymbol{\Omega}_1|\boldsymbol{\theta}, \boldsymbol{\alpha}, \mathbf{Y}_{\text{obs}}, \mathbf{W}_{\text{obs}}, \mathbf{U}_{\text{mis}}, \boldsymbol{\Omega}_2, \mathbf{V}, \mathbf{O})$, $p(\boldsymbol{\Omega}_2|\boldsymbol{\theta}, \boldsymbol{\alpha}, \mathbf{Y}_{\text{obs}}, \mathbf{W}_{\text{mis}}, \mathbf{U}_{\text{mis}}, \boldsymbol{\Omega}_1, \mathbf{V}, \mathbf{O})$, and $p(\boldsymbol{\alpha}, \mathbf{Y}_{\text{obs}}|\boldsymbol{\theta}, \mathbf{W}_{\text{obs}}, \mathbf{U}_{\text{mis}}, \boldsymbol{\Omega}_1, \boldsymbol{\Omega}_2, \mathbf{V}, \mathbf{O})$. As the implementation of the MH algorithm is similar to that given in Song and Lee (2004, 2005), it is not presented.

In the path sampling procedure, we augment the observed data $\mathbf{O}$ with the latent quantities $(\mathbf{Y}_{\text{obs}}, \mathbf{W}_{\text{obs}}, \mathbf{U}_{\text{mis}}, \boldsymbol{\Omega}_1, \boldsymbol{\Omega}_2, \mathbf{V})$ in the analysis. Consider the following class of densities defined by a continuous parameter t in [0, 1]:

$$p(\boldsymbol{\theta}, \boldsymbol{\alpha}, \mathbf{Y}_{\text{obs}}, \mathbf{W}_{\text{obs}}, \mathbf{U}_{\text{mis}}, \boldsymbol{\Omega}_1, \boldsymbol{\Omega}_2, \mathbf{V}|\mathbf{O}, t) = \frac{p(\boldsymbol{\theta}, \boldsymbol{\alpha}, \mathbf{Y}_{\text{obs}}, \mathbf{W}_{\text{obs}}, \mathbf{U}_{\text{mis}}, \boldsymbol{\Omega}_1, \boldsymbol{\Omega}_2, \mathbf{V}, \mathbf{O}|t)}{z(t)},$$

where $z(t) = p(\mathbf{O}|t)$. Recall that t is a parameter to link M_0 and M_1 such that for $a = 0, 1$, $z(a) = p(\mathbf{O}|t = a) = p(\mathbf{O}|M_a)$. Hence, $B_{10} = z(1)/z(0)$. It follows from the reasoning in Sect. 3 that

$$\log B_{10} = \frac{1}{2}\sum_{s=0}^{S}(t_{(s+1)} - t_{(s)})(\bar{\Delta}_{(s+1)} + \bar{\Delta}_{(s)}), \tag{23}$$

where $t_{(0)} = 0 < t_{(1)} < \cdots < t_{(S)} < t_{(S+1)} = 1$ are fixed grids in [0, 1], and

$$\bar{\Delta}_{(s)} = \frac{1}{J}\sum_{j=1}^{J} \Delta\left(\boldsymbol{\theta}^{(j)}, \boldsymbol{\alpha}^{(j)}, \mathbf{Y}_{\text{obs}}^{(j)}, \mathbf{W}_{\text{obs}}^{(j)}, \mathbf{U}_{\text{mis}}^{(j)}, \boldsymbol{\Omega}_1^{(j)}, \boldsymbol{\Omega}_2^{(j)}, \mathbf{V}^{(j)}, \mathbf{O}, t_{(s)}\right), \tag{24}$$

in which $\left\{\left(\boldsymbol{\theta}^{(j)}, \boldsymbol{\alpha}^{(j)}, \mathbf{Y}_{\text{obs}}^{(j)}, \mathbf{W}_{\text{obs}}^{(j)}, \mathbf{U}_{\text{mis}}^{(j)}, \boldsymbol{\Omega}_1^{(j)}, \boldsymbol{\Omega}_2^{(j)}, \mathbf{V}^{(j)}\right); j = 1, \ldots, J\right\}$ is a sample of observations that are simulated from $p(\boldsymbol{\theta}, \boldsymbol{\alpha}, \mathbf{Y}_{\text{obs}}, \mathbf{W}_{\text{obs}}, \mathbf{U}_{\text{mis}}, \boldsymbol{\Omega}_1, \boldsymbol{\Omega}_2, \mathbf{V}|\mathbf{O}, t)$, and

$$\begin{aligned}&\Delta(\boldsymbol{\theta}, \boldsymbol{\alpha}, \mathbf{Y}_{\text{obs}}, \mathbf{W}_{\text{obs}}, \mathbf{U}_{\text{mis}}, \boldsymbol{\Omega}_1, \boldsymbol{\Omega}_2, \mathbf{V}, \mathbf{O}, t)\\&= \frac{\mathrm{d}}{\mathrm{d}t} \log p(\mathbf{Y}_{\text{obs}}, \mathbf{W}_{\text{obs}}, \mathbf{U}_{\text{mis}}, \boldsymbol{\Omega}_1, \boldsymbol{\Omega}_2, \mathbf{V}, \mathbf{O}|\boldsymbol{\theta}, \boldsymbol{\alpha}, t).\end{aligned}$$

Note that $p(\mathbf{Y}_{\text{obs}}, \mathbf{W}_{\text{obs}}, \mathbf{U}_{\text{mis}}, \boldsymbol{\Omega}_1, \boldsymbol{\Omega}_2, \mathbf{V}, \mathbf{O}|\boldsymbol{\theta}, \boldsymbol{\alpha}, t)$ is the complete data likelihood that involves no integral.

5.3 An Illustrative Example

To illustrate the methodology, we analyze a small portion of the data set collected in the "Accelerated Schools for Quality Education (ASQE)" project, which was conducted by the Faculty of Education and the Hong Kong Institute of Educational Research, the Chinese University of Hong Kong. This project is an adaptation of the Accelerated Schools Projects initiated by Levin (see Levin, 1998) in the United States, with focus on a process of helping schools achieve an internal cultural change in order to be self-reliant in attaining school-based goals in self-improvement. Among the large number of objectives of this huge project, one particular issue is related with the "job satisfaction" of the teachers and (a) their "empowerment" to identify and solve the school's problems, and (b) the "school values inventory." These latent variables are important in the cultivation of their own and their peers' skills in improving their teaching skills and practice. The primary goal of our analysis is to apply a nonlinear structural equation model to assess the relationships of the mentioned latent variables. As the whole data set was collected from the principals, teachers, and students who were in $G = 50$ administrated schools, a two-level model is required. The three manifest variables that are served as indicators for the latent variable, "η, school value inventory" are (1) participation and collaboration, (2) collegiality, and (3) communication and consensus. These manifest variables are measured by the averages of seven, six, and ten items in the questionnaire. The corresponding Cronback's alpha on these scales are 0.93, 0.88, and 0.93, respectively. The three indicators for the latent factor, "ξ_1, teachers empowerment" are (1) decision making, (2) self efficacy, and (3) self autonomy, which are measured by the averages of four, four, and five items in the questionnaire. The Cronback's alpha are 0.79, 0.85 and 0.80, respectively. These manifest variables are treated as continuous. Three manifest variables (relating to questions: I proudly introduce my school as a worth-while working place to my friends; I find that my attitude of value is close to my school's attitude of value; and I can fully utilize my potentials in my school work) that are related with respondents' job satisfaction are taken as indicators for a latent factor, ξ_2, job satisfaction. These manifest variables are measured via a seven-point scale. To create dichotomous variables for illustration, data corresponding to these variables are transformed to $\{0, 1\}$ by grouping observations less than or equal to 4–0, and larger than 4–1. The teachers were also asked about their opinions on the impact of school effects, teacher effects, and student effects on the school improvement outcomes (SIO). These variables are measured by the average of four, seven, and four items in the questionnaire. To unify the scale, the continuous observations are standardized. After deleting a few observations with missing entries for brevity, the total number $N_1 + \cdots + N_{50}$ of observations is 1,555.

To demonstrate the impact of fixed covariates on the within-groups model, we include school effects on SIO as a fixed covariate c_{ugi} in the measurement equation and teacher effects on SIO as fixed covariates c_{1gi} in the structural equation. A two-level model with invariant parameters over groups and the following measurement equations is proposed:

$$M_0 : \mathbf{v}_g = \mathbf{\Lambda}_2\boldsymbol{\omega}_{2g} + \boldsymbol{\epsilon}_{2g}, \tag{25}$$

$$\mathbf{u}_{gi} = \mathbf{v}_g + \mathbf{A}_1 c_{ugi} + \mathbf{\Lambda}_1\boldsymbol{\omega}_{1gi} + \boldsymbol{\epsilon}_{1gi}, \tag{26}$$

where $\mathbf{c}_{ugi}$ is defined as above and $\mathbf{A}_1 = (a_1(1), \ldots, a_1(9))$. As the number of groups (G=50) is small, we do not consider a complicated model for the between-groups model. Thus, a simple factor analysis model with three correlated latent factors is used. Specifications of this model are

$$\mathbf{\Lambda}_2^T = \begin{bmatrix} 1^* & \lambda_2(2,1) & \lambda_2(3,1) & 0^* & 0^* & 0^* & 0^* & 0^* & 0^* \\ 0^* & 0^* & 0^* & 1^* & \lambda_2(5,2) & \lambda_2(6,2) & 0^* & 0^* & 0^* \\ 0^* & 0^* & 0^* & 0^* & 0^* & 0^* & 1^* & \lambda_2(8,3) & \lambda_2(9,3) \end{bmatrix},$$

and $diag(\mathbf{\Psi}_2) = (\psi_2(1), \ldots, \psi_2(9))$, where parameters with an asterisk are fixed. The unknown parameters in $\mathbf{\Phi}_2$ are denoted by $\{\phi_2(1,1), \phi_2(1,2), \phi_2(1,3),$ $\phi_2(2,2), \phi_2(2,3), \phi_2(3,3)\}$. For the within-groups model, the factor loading matrix $\mathbf{\Lambda}_1 = (\lambda_1(k,h))$ corresponding to latent factors $(\eta_{1gi}, \xi_{1gi1}, \xi_{1gi2})$ is taken to have the same form as $\mathbf{\Lambda}_2$, and $diag(\mathbf{\Psi}_1) = (\psi_1(1), \ldots, \psi_1(9))$, where $\psi_1(7)$, $\psi_1(8)$, and $\psi_1(9)$ that are corresponding to dichotomous manifest variables are fixed at 1.0 for identification purpose. We consider the following nonlinear structural equation to investigate effects of the fixed covariates c_{1gi} and the interaction and quadratic terms of ξ_{1gi1} and ξ_{1gi2}:

$$\begin{aligned}\eta_{1gi} &= b_1 c_{1gi} + \gamma_1(1)\xi_{1gi1} + \gamma_1(2)\xi_{1gi2} + \gamma_1(3)\xi_{1gi1}^2 + \gamma_1(4)\xi_{1gi1}\xi_{1gi2} \\ &\quad + \gamma_1(5)\xi_{1gi2}^2 + \delta_{1gi},\end{aligned} \tag{27}$$

where c_{1gi} is defined as above. The unknown parameters in $\boldsymbol{\alpha}$, $\mathbf{\Gamma}_1$, and $\mathbf{\Phi}_1$ are denoted by $\{\alpha_1, \alpha_2, \alpha_3\}$ (that corresponds to the thresholds of the three dichotomous variables), $\{\gamma_1(1), \gamma_1(2), \gamma_1(3), \gamma_1(4), \gamma_1(5)\}$, and $\{\phi_1(1,1), \phi_1(1,2)\ \phi_1(2,2)\}$, respectively.

In this illustrative example, we have little prior information about the data. Thus, some data-dependent prior inputs were used for the hyper-parameter values in the conjugate prior distributions. These prior inputs are obtained by conducting an auxiliary Bayesian estimation with proper vague conjugate prior distributions, which gives estimates $\tilde{\mathbf{A}}_{1k}$, $\tilde{\mathbf{\Lambda}}_{1k}$, $\tilde{\mathbf{\Lambda}}_{1\omega k}$, and $\tilde{\mathbf{\Lambda}}_{2k}$, where $\mathbf{A}_{1k}$, $\mathbf{\Lambda}_{1k}$, $\mathbf{\Lambda}_{1\omega k}$, $\mathbf{\Lambda}_{2k}$, and below hyper-parameters are similarly defined as in (16), see more details in Appendix. Some less important hyper-parameter values are fixed at some ad hoc prior inputs as below: $\mathbf{H}_{01yk}$, $\mathbf{H}_{01\omega k}$, $\mathbf{H}_{02yk}$, $\mathbf{H}_{01k}$, and $\mathbf{H}_{02k}$ are fixed at identity matrices with appropriate dimensions, $\mathbf{R}_{01}$ and $\mathbf{R}_{02}$ are fixed at $5\mathbf{I}$. Other kinds of data-dependence have been proposed in Bayesian model selection (see, for example, Raftery, 1996; Richardson and Green, 1997, among others). We emphasize that the data-dependent prior inputs are used for the purpose of illustration only in this example; we are not routinely recommending this method for substantive practical applications.

To study the sensitivity of the results with respect to different prior informations, we consider the following three types of prior inputs (see (37) and (39) in Appendix):

(I) $\alpha_{01\epsilon k} = \alpha_{02\epsilon k} = \alpha_{01\delta k} = 3$, $\beta_{01\epsilon k} = \beta_{02\epsilon k} = \beta_{01\delta k} = 10$, $\rho_{01} = \rho_{02} = 8$, $\mathbf{\Lambda}_{01k} = \tilde{\mathbf{\Lambda}}_{1k}$, $\mathbf{\Lambda}_{01k} = \tilde{\mathbf{\Lambda}}_{1k}$, $\mathbf{\Lambda}_{01\omega k} = \tilde{\mathbf{\Lambda}}_{1\omega k}$, and $\mathbf{\Lambda}_{02k} = \tilde{\mathbf{\Lambda}}_{2k}$.

(II) $\alpha_{01\epsilon k} = \alpha_{02\epsilon k} = \alpha_{01\delta k} = 3$, $\beta_{01\epsilon k} = \beta_{02\epsilon k} = \beta_{01\delta k} = 5$, $\rho_{01} = \rho_{02} = 6$, the other specified hyper-parameters are equal to half of those values given in (I).

(III) $\alpha_{01\epsilon k} = \alpha_{02\epsilon k} = \alpha_{01\delta k} = 3$, $\beta_{01\epsilon k} = \beta_{02\epsilon k} = \beta_{01\delta k} = 15$, $\rho_{01} = \rho_{02} = 10$, the other specified hyper-parameters are equal to twice of those values given in (I).

Let M_0 be the model that is defined by (25), (26), and (27) with the between-groups factor analysis model. The path sampling procedure is applied to compute the Bayes factor for comparing the M_0 with the following models:

M_1 : Between-groups model is defined by (25), with a linear structural equation,

$$\eta_{2g} = \gamma_2(1)\xi_{2g1} + \gamma_2(2)\xi_{2g2} + \delta_{2g}, \tag{28}$$

rather than the factor model. Within-groups model is defined by (26) and (27).

M_2 : Between-groups model is defined by (25). Within-groups model is defined by (26) and the following linear structural equation:

$$\eta_{1gi} = b_1 c_{1gi} + \gamma_1(1)\xi_{1gi1} + \gamma_1(2)\xi_{1gi2} + \delta_{1gi}. \tag{29}$$

M_3 : Between-groups model is defined by (25). Within-groups model is defined by (26) and the following nonlinear structural equation without fixed covariate:

$$\begin{aligned}\eta_{1gi} = {} & \gamma_1(1)\xi_{1gi1} + \gamma_1(2)\xi_{1gi2} + \gamma_1(3)\xi^2_{1gi1} + \gamma_1(4)\xi_{1gi1}\xi_{1gi2} \\ & +\gamma_1(5)\xi^2_{1gi2} + \delta_{1gi}.\end{aligned} \tag{30}$$

M_4 : Between-groups model is defined by (25). Within-groups model is defined by (27) and the following measurement equation in relation to $\mathbf{u}_{gi}$ without any fixed covariate: $\mathbf{u}_{gi} = \mathbf{v}_g + \mathbf{\Lambda}_1\boldsymbol{\omega}_{1gi} + \boldsymbol{\epsilon}_{1gi}$.

The path sampling procedure is applied to compute the Bayes factors for comparing models $M_0, \ldots, M_4$, under priors inputs (I), (II), and (III). From (23) and (24), we see that the integral part in the computation is the simulation of the random sample $\{(\boldsymbol{\theta}^{(j)}, \boldsymbol{\alpha}^{(j)}, \mathbf{Y}^{(j)}_{\text{obs}}, \mathbf{W}^{(j)}_{\text{obs}}, \mathbf{U}^{(j)}_{\text{mis}}, \mathbf{\Omega}^{(j)}_1, \mathbf{\Omega}^{(j)}_2, \mathbf{V}^{(j)}); j = 1, \ldots, J\}$ by means of the posterior simulation via the MCMC methods. In the computation of the Bayes factor, we take $S = 10$, and further collect $J = 3,000$ iterations after convergence

Table 4 Bayes factor estimates for model comparison: ASQE data set

Logarithm of Bayes factor	Prior I	Prior II	Prior III
$\log B_{01}$	16.05	9.31	18.11
$\log B_{02}$	3.43	4.02	3.16
$\log B_{03}$	5.82	5.71	4.99
$\log B_{04}$	89.34	91.13	90.89

at each grid. The computed Bayes factors under different prior inputs are reported in Table 4. We note that the logarithm Bayes factors that are obtained under the different choices of prior informations are close. As all $\log B_{01}, \ldots, \log B_{04}$ are larger than 3.0, they suggest the same conclusion that M_0 is the best model. To cross validate, we have computed the Bayes factors with larger S and J and obtain the same conclusion as above. Based on the model comparison result on the nonnested models M_0 and M_1, we find that in fitting the between-groups model a factor analysis model is better than a SEM with a linear structural equation that is defined by (28). Hence, we conclude that the data give support of evidence to correlated between-groups level latent factors, rather than the relationships among these latent factors that are given by (28). Based on the results of model comparison of M_0 with M_2 and M_3, we conclude that at the within-groups level, the fixed covariate c_{1gi} (teacher effects on school improvement outcomes) and the nonlinear terms of ξ_{1gi1} (teacher empowerment) and ξ_{1gi2} (teacher's job satisfaction) have substantial effects on the endogenous latent variable η_{1gi} (school value inventory). As M_0 is better than M_4, we conclude that for the measurement equation of the within-groups model, the effect of fixed covariate c_{ugi} (school effects on SIO) is significant. Based on these findings, we conclude that at the within-group level, the school effects on school improvement outcomes are important in relating the manifest variables with their latent factors. The above results on model comparison and conclusion cannot be achieved without a comprehensive model that combines the individual multilevel SEM, NSEM, SEM with fixed covariates, and SEM with dichotomous and ordered categorical variables together.

Under the three types of prior inputs (I), (II), and (III), Bayesian estimates of the parameters under M_0 are obtained by the algorithm that combines the Gibbs sampler and the MH algorithm. Here, we collect $T = 4,000$ observations after the burn-in phase of 4,000 observations to produce the Bayesian estimates, their standard error estimates, and the 95% HPD intervals. Bayesian estimates obtained under different priors are close. This agrees with the common understanding that Bayesian estimation is not sensitive to prior inputs. To save space, only results obtained under prior inputs (I) are reported in Table 5. The PP p-value (Gelman et al., 1996) is equal to 0.59, which indicates that the proposed model fits the sample data. Most of the 95% HPD intervals are reasonably short. Comparatively, the standard errors and HPD intervals are larger for estimates of the between-groups parameters. This indicates that as G is comparatively small, the variability of the estimates of the

Table 5 Bayesian estimates, their standard error estimates, and HPD intervals under Prior I: ASQE data set

PAR	EST	STD	HPD	PAR	EST	STD	HPD
$\lambda_1(2,1)$	0.957	0.020	[0.916, 0.999]	$\lambda_2(2,1)$	0.275	0.263	[−0.301, 0.796]
$\lambda_1(3,1)$	0.894	0.021	[0.849, 0.938]	$\lambda_2(3,1)$	0.291	0.269	[−0.299, 0.822]
$\lambda_1(5,2)$	1.015	0.046	[0.916, 1.114]	$\lambda_2(5,2)$	0.051	0.263	[−0.526, 0.587]
$\lambda_1(6,2)$	0.723	0.045	[0.622, 0.806]	$\lambda_2(6,2)$	0.104	0.318	[−0.561, 0.751]
$\lambda_1(8,3)$	1.138	0.148	[0.877, 2.043]	$\lambda_2(8,3)$	0.583	0.351	[−0.078, 1.478]
$\lambda_1(9,3)$	0.536	0.067	[0.399, 0.778]	$\lambda_2(9,3)$	0.242	0.254	[−0.249, 0.829]
$\phi_1(1,1)$	0.392	0.028	[0.338, 0.459]	$\phi_2(1,1)$	0.184	0.054	[0.094, 0.298]
$\phi_1(1,2)$	0.568	0.052	[0.386, 0.682]	$\phi_2(1,2)$	0.009	0.036	[−0.065, 0.083]
$\phi_1(2,2)$	2.098	0.346	[1.038, 2.921]	$\phi_2(1,3)$	0.024	0.051	[0.088, 0.277]
$\psi_1(1)$	0.107	0.006	[0.095, 0.119]	$\phi_2(2,2)$	0.172	0.049	[−0.078, 0.138
$\psi_1(2)$	0.156	0.007	[0.141, 0.171]	$\phi_2(2,3)$	0.012	0.050	[−0.093, 0.124]
$\psi_1(3)$	0.179	0.008	[0.163, 0.197]	$\phi_2(3,3)$	0.300	0.108	[0.119, 0.556]
$\psi_1(4)$	0.442	0.022	[0.363, 0.486]	$\psi_2(1)$	0.525	0.113	[0.325, 0.769]
$\psi_1(5)$	0.399	0.022	[0.348, 0.442]	$\psi_2(2)$	0.441	0.089	[0.268, 0.624]
$\psi_1(6)$	0.538	0.023	[0.494, 0.588]	$\psi_2(3)$	0.443	0.088	[0.291, 0.646]
b_1	0.194	0.030	[0.116, 0.270]	$\psi_2(4)$	0.516	0.110	[0.331, 0.757]
$\gamma_1(1)$	0.234	0.058	[0.111, 0.337]	$\psi_2(5)$	0.423	0.084	[0.274, 0.605]
$\gamma_1(2)$	0.148	0.028	[0.096, 0.252]	$\psi_2(6)$	0.511	0.101	[0.328, 0.734]
$\gamma_1(3)$	0.507	0.098	[−0.006, 0.694]	$\psi_2(7)$	0.859	0.204	[0.461, 1.257]
$\gamma_1(4)$	−0.394	0.090	[−0.635, 0.023]	$\psi_2(8)$	0.780	0.179	[0.448, 1.219]
$\gamma_1(5)$	0.050	0.025	[−0.060, 0.120]	$\psi_2(9)$	0.572	0.120	[0.354, 0.822]
$a_1(1)$	0.412	0.028	[0.330, 0.482]	α_1	−0.109	0.162	[−0.440, 0.212]
$a_1(2)$	0.418	0.027	[0.334, 0.480]	α_2	0.841	0.167	[0.518, 1.293]
$a_1(3)$	0.469	0.026	[0.394, 0.531]	α_3	−0.341	0.119	[−0.579, −0.092]
$a_1(4)$	0.390	0.025	[0.334, 0.438]				
$a_1(5)$	0.443	0.024	[0.385, 0.487]				
$a_1(6)$	0.412	0.023	[0.360, 0.455]				
$a_1(7)$	1.479	0.123	[1.122, 1.724]				
$a_1(8)$	1.548	0.163	[1.242, 1.995]				
$a_1(9)$	0.785	0.057	[0.653, 0.892]				
$\psi_{1\delta}$	0.310	0.017	[0.276, 0.360]				

between-groups parameters is larger. We also use the following estimated residuals to reveal the adequacy of the proposed measurement model and structural equation for fitting the data: $\hat{\boldsymbol{\epsilon}}_{1gi} = \mathbf{u}_{gi} - \hat{\mathbf{v}}_g - \hat{\mathbf{A}}_1 c_{ugi} - \hat{\boldsymbol{\Lambda}}_1 \hat{\boldsymbol{\omega}}_{1gi}$, $\hat{\boldsymbol{\epsilon}}_{2g} = \hat{\mathbf{v}}_g - \hat{\boldsymbol{\Lambda}}_2 \hat{\boldsymbol{\omega}}_{2g}$, and $\hat{\delta}_{1gi} = \hat{\eta}_{1gi} - \hat{\gamma}_1(1)\hat{\xi}_{1gi1} - \hat{\gamma}_1(2)\hat{\xi}_{1gi2} - \hat{\gamma}_1(3)\hat{\xi}^2_{1gi1} - \hat{\gamma}_1(4)\hat{\xi}_{1gi1}\hat{\xi}_{1gi2} - \hat{\gamma}_1(5)\hat{\xi}^2_{1gi2}$.

Plots of the estimated residual $\hat{\epsilon}_{1gi1}$ vs. $\hat{\xi}_{1gi1}$, $\hat{\delta}_{1gi}$ vs. $\hat{\xi}_{1gi1}$, and $\hat{\epsilon}_{2g1}$ vs. $\hat{\xi}_{2g1}$ are presented in Fig. 1. Other estimated residual plots have similar behaviors. These plots lie within two parallel horizontal lines that are centered at zero, and no linear or quadratic trends are detected. This roughly indicates that the proposed first-level measurement model and structural equation and the second-level measurement model are adequate in fitting the data.

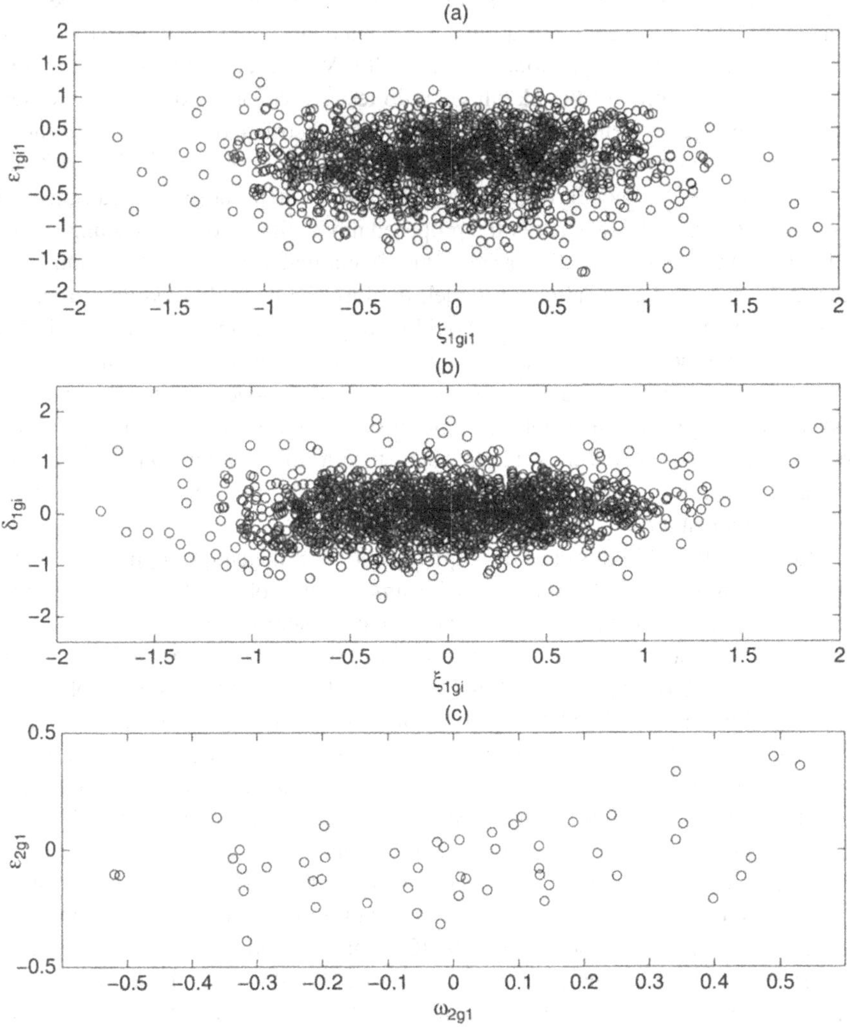

Fig. 1 (**a**) Plot of $\hat{\xi}_{1gi1}$ vs. $\hat{\epsilon}_{1gi1}$; (**b**) Plot of $\hat{\xi}_{1gi1}$ vs. $\hat{\delta}_{1gi}$; (**c**) Plot of $\hat{\xi}_{2g1}$ vs. $\hat{\epsilon}_{2g1}$

6 Discussion

In structural equation modeling or covariance structure analysis, an important issue is goodness-of-fit test of the hypothesized model. The traditional approach is based on the likelihood ratio test statistic that can be proved to have an asymptotic chi-square distribution. In certain sense, its emphasis is on comparing the hypothesized model with a so-called "saturated" model, which is commonly regarded as a model with an unstructured covariance matrix but under the normality distribution. In practical applications, it is hoped to have a hypothesized model that is not rejected, so that it can be used for supporting some theory, etc. When dealing with some nonstandard SEMs, this traditional approach will encounter the following two difficulties, in addition to those mentioned in Sect. 1: (a) Very often, the basic assumptions required by the theory of the likelihood ratio test are violated, for example, observations may not be identically and independently distributed, so that it is difficult to derive the asymptotic distribution of the likelihood ratio test. (b) It is difficult to define a saturated model. Rigorously speaking, if the purpose of goodness-of-fit assessment is to evaluate whether the proposed model fits the observed sample data, a model with an unrestricted covariance matrix but under the normality assumption cannot be regarded as a saturated model, because the normality assumption is restrictive; rather, a saturated model should be a general nonparametric model with essentially no model or distributional assumption. However, it is very difficult for the tradition approach to deal with this sort of saturated model with no distributional assumption. Moreover, it is not desirable to stop trying to improve the model just because it is not rejected by some test statistics. Model comparison statistics like the Bayes factor are useful to search for a better model in fitting the given data in practical applications.

As we explained in this chapter, implementation of the path sampling procedure is conceptually simple; it just requires the simulation of observations from the posterior distribution. Thanks for the recent advance of statistical computing, there are many efficient tools for computing the task. In addition to the examples given in this chapter, Lee (2007) demonstrated that the Bayes factor and the path sampling can be applied to other useful SEMs, such as multisample SEMs, nonlinear SEMs with missing data that are missing with an nonignorable mechanism, nonlinear SEMs with manifest variables from an exponential family distribution, among others. In applying path sampling to compute Bayes factor for model comparison of SEMs, the values of S and J involved in (11) and (12) have to be chosen. For most situations, the choice of J depends on the complexity of the model, and the nature of the data. Based on our experience in analyzing dichotomous data, it requires more iterations to achieve convergence in the MCMC algorithm, and requires more simulated observations to approximate the various statistics. For this kind of data, we suggest that J should be equal to or larger than 3,000. From the criterion given in (1), the conclusion based on a value of $2\log B_{10}$ that is less than 6 is not strong. Under such case, it is worthwhile to compute the Bayes factor with larger values of S and J to make sure the approximation is good; moreover, the result should be cross-validated by other diagnostic checks, such as estimated residual plots that can be conveniently obtained through the Bayesian estimates of the parameters and latent variables (see Lee, 2007).

Gelman and Meng (1998) pointed out that the path sampling can be applied to any two completing models with the same support. In this chapter, we have shown that path sampling can be flexibly applied to some complex SEMs. However, for some models, for instance, robust SEM with stochastic weights in the covariance matrixes of the latent variables or/error measurements (see Lee and Xia, 2006) or nonparametric SEMs with general distributional assumptions, it is either difficult to apply the path sampling, or the results are not satisfactory. Further research to cope with these complicated situations is necessary.

Acknowledgments This research is fully supported by two grants (CUHK 404507 and 450607) from the Research Grant Council of the Hong Kong Special Administrative Region, and a direct grant from the Chinese University of Hong Kong (Project ID 2060278). The authors are indebted to Dr. John C. K. Lee, Faculty of Education, The Chinese University of Hong Kong, for providing the data in the application.

Appendix: Full Conditional Distributions

The conditional distributions required by the Gibbs sampler in the posterior simulation of the integrated model will be presented in this appendix. We use $p(\cdot|\cdot)$ to denote the conditional distribution if the context is clear, and note that $(\mathbf{Y}_{\text{obs}}, \mathbf{W}_{\text{obs}}, \mathbf{U}_{\text{mis}}, \mathbf{O}) = \mathbf{U}$.

(i) $p(\mathbf{V}|\boldsymbol{\theta}, \boldsymbol{\alpha}, \mathbf{Y}_{\text{obs}}, \mathbf{W}_{\text{obs}}, \mathbf{U}_{\text{mis}}, \boldsymbol{\Omega}_1, \boldsymbol{\Omega}_2, \mathbf{O}) = p(\mathbf{V}|\boldsymbol{\theta}, \mathbf{U}, \boldsymbol{\Omega}_1, \boldsymbol{\Omega}_2)$: This conditional distribution is equal to a product of $p(\mathbf{v}_g|\boldsymbol{\theta}, \mathbf{U}_g, \boldsymbol{\Omega}_{1g}, \boldsymbol{\omega}_{2g})$ with $g = 1, \ldots, G$. For each gth term in this product, its conditional distribution is $N[\boldsymbol{\mu}_g^*, \boldsymbol{\Sigma}_g^*]$, where

$$\boldsymbol{\mu}_g^* = \boldsymbol{\Sigma}_g^* \left\{ \boldsymbol{\Psi}_1^{-1} \sum_{i=1}^{N_g} [\mathbf{u}_{gi} - \mathbf{A}_1 \mathbf{c}_{ugi} - \boldsymbol{\Lambda}_1 \boldsymbol{\omega}_{1gi}] + \boldsymbol{\Psi}_2^{-1} [\mathbf{A}_2 \mathbf{c}_{vg} + \boldsymbol{\Lambda}_2 \boldsymbol{\omega}_{2g}] \right\}, \quad \text{and}$$

$$\boldsymbol{\Sigma}_g^* = (N_g \boldsymbol{\Psi}_1^{-1} + \boldsymbol{\Psi}_2^{-1})^{-1}. \tag{31}$$

(ii) $p(\boldsymbol{\Omega}_1|\boldsymbol{\theta}, \boldsymbol{\alpha}, \mathbf{Y}_{\text{obs}}, \mathbf{W}_{\text{obs}}, \mathbf{U}_{\text{mis}}, \boldsymbol{\Omega}_2, \mathbf{V}, \mathbf{O}) = p(\boldsymbol{\Omega}_1|\boldsymbol{\theta}, \mathbf{U}, \boldsymbol{\Omega}_2, \mathbf{V}) = \prod_{g=1}^{G} \prod_{i=1}^{N_g} p(\boldsymbol{\omega}_{1gi}|\boldsymbol{\theta}, \mathbf{v}_g, \boldsymbol{\omega}_{2g}, \mathbf{u}_{gi})$, where $p(\boldsymbol{\omega}_{1gi}|\boldsymbol{\theta}, \mathbf{v}_g, \boldsymbol{\omega}_{2g}, \mathbf{u}_{gi})$ is proportional to

$$\begin{aligned}
\exp\Bigg[-\frac{1}{2} \Big\{ & \boldsymbol{\xi}_{1gi}^T \boldsymbol{\Phi}_1^{-1} \boldsymbol{\xi}_{1gi} + [\mathbf{u}_{gi} - \mathbf{v}_g - \mathbf{A}_1 \mathbf{c}_{ugi} - \boldsymbol{\Lambda}_1 \boldsymbol{\omega}_{1gi}]^T \boldsymbol{\Psi}_1^{-1} \\
& \times [\mathbf{u}_{gi} - \mathbf{v}_g - \mathbf{A}_1 \mathbf{c}_{ugi} - \boldsymbol{\Lambda}_1 \boldsymbol{\omega}_{1gi}] + [\boldsymbol{\eta}_{1gi} - \mathbf{B}_1 \mathbf{c}_{1gi} - \boldsymbol{\Pi}_1 \boldsymbol{\eta}_{1gi} - \boldsymbol{\Gamma}_1 \mathbf{F}_1(\boldsymbol{\xi}_{1gi})]^T \\
& \times \boldsymbol{\Psi}_{1\delta}^{-1} [\boldsymbol{\eta}_{1gi} - \mathbf{B}_1 \mathbf{c}_{1gi} - \boldsymbol{\Pi}_1 \boldsymbol{\eta}_{1gi} - \boldsymbol{\Gamma}_1 \mathbf{F}_1(\boldsymbol{\xi}_{1gi})] \Big\} \Bigg].
\end{aligned} \tag{32}$$

(iii) $p(\boldsymbol{\Omega}_2|\boldsymbol{\theta}, \boldsymbol{\alpha}, \mathbf{Y}_{\text{obs}}, \mathbf{W}_{\text{obs}}, \mathbf{U}_{\text{mis}}, \boldsymbol{\Omega}_1, \mathbf{V}, \mathbf{O})$: This distribution has very similar form as in $p(\boldsymbol{\Omega}_1|\cdot)$ and (32), hence is not presented to save space.

(iv) $p(\boldsymbol{\alpha}, \mathbf{Y}_{\text{obs}}|\boldsymbol{\theta}, \mathbf{W}_{\text{obs}}, \mathbf{U}_{\text{mis}}, \boldsymbol{\Omega}_1, \boldsymbol{\Omega}_2, \mathbf{V}, \mathbf{O})$: To deal with the situation with little or no information about these parameters, the following noninformation prior distribution is used: $p(\boldsymbol{\alpha}_k) = p(\alpha_{k,2}, \ldots, \alpha_{k,b_k-1}) \propto C, \ k = 1, \ldots, s$, where C is a constant. As $(\boldsymbol{\alpha}, \mathbf{Y}_g)$ is independent with $(\boldsymbol{\alpha}, \mathbf{Y}_h)$ for $g \neq h$, and that $\boldsymbol{\Psi}_1$ is diagonal, we have

$$p(\boldsymbol{\alpha}, \mathbf{Y}|\cdot) = \prod_{g=1}^{G} p(\boldsymbol{\alpha}_h, \mathbf{Y}_g|\cdot) = \prod_{g=1}^{G} \prod_{k=1}^{s} p(\boldsymbol{\alpha}_k, \mathbf{Y}_{gk}|\cdot), \tag{33}$$

where $\mathbf{Y}_{gk} = [y_{gk1}, \ldots, y_{gkN_g}]$. Let ψ_{1k} be the kth diagonal element of $\boldsymbol{\Psi}_1$, v_{gk} be the kth element of $\mathbf{v}_g$, and $\boldsymbol{\Lambda}_{1k}$ be the kth row of $\boldsymbol{\Lambda}_1$, and $I_A(y)$ be an indicator function with value 1 if y is A and zero otherwise, $p(\boldsymbol{\alpha}, \mathbf{Y}|\cdot)$ can be obtained from (33) and

$$p(\boldsymbol{\alpha}_k, y_{gki}|\cdot) \propto \prod_{i=1}^{N_g} \Phi^*\left\{\psi_{1k}^{-1/2}[y_{gki} - v_{gk} - \boldsymbol{\Lambda}_{1k}^T \boldsymbol{\omega}_{1gi}]\right\} I_{(\alpha_{k,z_{gki}}, \alpha_{k,z_{gki}+1}]}(y_{gki}). \tag{34}$$

(v) $p(\mathbf{W}_{\text{obs}}|\boldsymbol{\theta}, \boldsymbol{\alpha}, \mathbf{Y}_{\text{obs}}, \mathbf{U}_{\text{mis}}, \boldsymbol{\Omega}_1, \boldsymbol{\Omega}_2, \mathbf{V}, \mathbf{O}) = \prod_{k=1}^{s} \prod_{g=1}^{G} \prod_{i=1}^{n_{g,k}} p(w_{gik,\text{obs}}|\boldsymbol{\theta}, \boldsymbol{\omega}_{1gi}, d_{gik,\text{obs}})$. Moreover, it follows from the definition of the model that

$$p(w_{gik,\text{obs}}|\boldsymbol{\theta}, \omega_{1gi}, d_{gik,\text{obs}}) \sim \begin{cases} N[\boldsymbol{\Lambda}_{1k}' \omega_{1gi}, \psi_{1k}] I_{(-\infty,0)}(w_{gik,\text{obs}}), & \text{if } d_{gik,\text{obs}} = 0 \\ N[\boldsymbol{\Lambda}_{1k}' \omega_{1gi}, \psi_{1k}] I_{(0,\infty)}(w_{gik,\text{obs}}), & \text{if } d_{gik,\text{obs}} = 1, \end{cases} \tag{35}$$

where $n_{g,k}$ is the number of $d_{gki,\text{obs}}$ in $\mathbf{D}_{k,\text{obs}}$ and $\mathbf{D}_{k,\text{obs}}$ is the kth row of $\mathbf{D}_{\text{obs}}$.

(vi) $p(\mathbf{U}_{\text{mis}}|\boldsymbol{\theta}, \boldsymbol{\alpha}, \mathbf{Y}_{\text{obs}}, \mathbf{W}_{\text{obs}}, \boldsymbol{\Omega}_1, \boldsymbol{\Omega}_2, \mathbf{V}, \mathbf{O}) = \prod_{g=1}^{G} \prod_{i=1}^{N_g} p(\mathbf{u}_{gi,\text{mis}}|\boldsymbol{\theta}, \boldsymbol{\omega}_{1gi}, \mathbf{v}_g)$, and

$$[\mathbf{u}_{gi,\text{mis}}|\boldsymbol{\theta}, \boldsymbol{\omega}_{1gi}] \overset{D}{=} N[\mathbf{v}_g + \mathbf{A}_{1i,\text{mis}} \mathbf{c}_{ugi} + \boldsymbol{\Lambda}_{1i,\text{mis}} \boldsymbol{\omega}_{1gi}, \boldsymbol{\Psi}_{1i,\text{mis}}], \tag{36}$$

where $\mathbf{A}_{1i,\text{mis}}$ and $\boldsymbol{\Lambda}_{1i,\text{mis}}$ are submatrices of $\mathbf{A}_1$ and $\boldsymbol{\Lambda}_1$ with rows that correspond to observed components deleted, and $\boldsymbol{\Psi}_{1i,\text{mis}}$ is a submatrix of $\boldsymbol{\Psi}_1$ with the appropriate rows and columns deleted.

(vii) $p(\boldsymbol{\theta}|\boldsymbol{\alpha}, \mathbf{Y}_{\text{obs}}, \mathbf{W}_{\text{obs}}, \mathbf{U}_{\text{mis}}, \boldsymbol{\Omega}_1, \boldsymbol{\Omega}_2, \mathbf{V}, \mathbf{O}) = p(\boldsymbol{\theta}|\mathbf{U}, \boldsymbol{\Omega}_1, \boldsymbol{\Omega}_2)$: Let $\boldsymbol{\theta}_1$ be the vector of unknown parameters in $\mathbf{A}_1$, $\boldsymbol{\Lambda}_1$, and $\boldsymbol{\Psi}_1$, $\boldsymbol{\theta}_{1\omega}$ be the vector of unknown parameters in $\boldsymbol{\Pi}_1$, $\boldsymbol{\Gamma}_1$, $\boldsymbol{\Phi}_1$, and $\boldsymbol{\Psi}_{1\delta}$, $\boldsymbol{\theta}_2$ be the vector of unknown parameters in $\mathbf{A}_2$, $\boldsymbol{\Lambda}_2$, and $\boldsymbol{\Psi}_2$, and $\boldsymbol{\theta}_{2\omega}$ be the vector of unknown parameters in $\boldsymbol{\Pi}_2$, $\boldsymbol{\Gamma}_2$, $\boldsymbol{\Phi}_2$, and $\boldsymbol{\Psi}_{2\delta}$.

For $\boldsymbol{\theta}_1$, the following commonly used conjugate type prior distributions are used:

$$p(\psi_{1k}^{-1}) \overset{D}{=} Gamma[\alpha_{01\epsilon k}, \beta_{01\epsilon k}], \quad p(\boldsymbol{\Lambda}_{1k}|\psi_{1k}) \overset{D}{=} N[\boldsymbol{\Lambda}_{01k}, \psi_{1k}\mathbf{H}_{01yk}],$$
$$p(\mathbf{A}_{1k}) \overset{D}{=} N[\mathbf{A}_{01k}, \mathbf{H}_{01k}], \quad k = 1, \ldots, p, \tag{37}$$

where $\mathbf{A}_{1k}^T$, $\boldsymbol{\Lambda}_{1k}^T$ are the row vectors that contain the unknown parameters in the kth row of $\mathbf{A}_1$ and $\boldsymbol{\Lambda}_1$, respectively; $\alpha_{01\epsilon k}$, $\beta_{01\epsilon k}$, $\mathbf{A}_{01k}$, $\boldsymbol{\Lambda}_{01k}$, $\mathbf{H}_{01k}$, and $\mathbf{H}_{01yk}$ are given hyper-parameters values. For $k \neq h$, it is assumed that $(\psi_{1k}, \boldsymbol{\Lambda}_{1k})$ and $(\psi_{1h}, \boldsymbol{\Lambda}_{1h})$ are independent. Let $\mathbf{U}^* = \{\mathbf{u}_{gi} - \mathbf{v}_g - \mathbf{A}_1\mathbf{c}_{ugi};\ i = 1, \ldots, N_g,\ g = 1, \ldots, G\}$ and $\mathbf{U}_k^{*^T}$ be the kth row of $\mathbf{U}^*$, $\boldsymbol{\Omega}_1 = \{\boldsymbol{\omega}_{1gi};\ g = 1, \ldots, G, i = 1, \ldots, N_g\}$, $\Sigma_{1k} = (\mathbf{H}_{01yk}^{-1} + \boldsymbol{\Omega}_1\boldsymbol{\Omega}_1^T)^{-1}$, $\mathbf{m}_{1k} = \Sigma_{1k}(\mathbf{H}_{01yk}^{-1}\boldsymbol{\Lambda}_{01k} + \boldsymbol{\Omega}_1\mathbf{U}_k^*)$, and $\beta_{1\epsilon k} = \beta_{01\epsilon k} + \frac{1}{2}(\mathbf{U}_k^{*^T}\mathbf{U}_k^* - \mathbf{m}_{1k}^T\boldsymbol{\Omega}_1^{-1}\mathbf{m}_{1k} + \boldsymbol{\Lambda}_{01k}^T\mathbf{H}_{01yk}^{-1}\boldsymbol{\Lambda}_{01k})$, it can be shown that

$$p(\psi_{1k}^{-1}|\cdot) \overset{D}{=} \text{Gamma}\ (2^{-1}N_g + \alpha_{01\epsilon k}, \beta_{1\epsilon k}), \quad \text{and} \quad p(\boldsymbol{\Lambda}_{1k}|\psi_{1k}, \cdot) \overset{D}{=} N[\mathbf{m}_{1k}, \psi_{1k}\Sigma_{1k}]. \tag{38}$$

Let $\mathbf{C}_u = \{\mathbf{c}_{ugi};\ i = 1, \ldots, N_g,\ g = 1, \ldots, G\}$, $\tilde{\mathbf{U}} = \{\mathbf{u}_{gi} - \mathbf{v}_g - \mathbf{\Lambda}_1\boldsymbol{\omega}_{1gi};\ i = 1, \ldots, N_g,\ g = 1, \ldots, G\}$, and $\tilde{\mathbf{U}}_k^T$ be the kth row of $\tilde{\mathbf{U}}$, $\tilde{\Sigma}_{1k} = (\mathbf{H}_{01k}^{-1} + \mathbf{C}_u\mathbf{C}_u^T)^{-1}$, $\tilde{\mathbf{m}}_{1k} = \tilde{\Sigma}_{1k}(\mathbf{H}_{01k}^{-1}\mathbf{A}_{01k} + \mathbf{C}_u\tilde{\mathbf{U}}_k)$.

For $\boldsymbol{\theta}_{1\omega}$, it is assumed that $\mathbf{\Phi}_1$ is independent of $(\mathbf{\Lambda}_{1\omega}, \mathbf{\Psi}_{1\delta})$, where $\mathbf{\Lambda}_{1\omega} = (\mathbf{B}_1^T, \mathbf{\Pi}_1^T, \mathbf{\Gamma}_1^T)^T$. Also, $(\mathbf{\Lambda}_{1\omega k}, \psi_{1\delta k})$ and $(\mathbf{\Lambda}_{1\omega h}, \psi_{1\delta h})$ are independent, where $\mathbf{\Lambda}_{1\omega k}$ and $\psi_{1\delta k}$ are the kth row and diagonal element of $\mathbf{\Lambda}_{1\omega}$ and $\mathbf{\Psi}_{1\delta}$, respectively. The associated prior distribution of $\mathbf{\Phi}_1$ is $p(\mathbf{\Phi}_1^{-1}) \stackrel{D}{=} W[\mathbf{R}_{01}, \rho_{01}, q_{12}]$, where $W[\cdot, \cdot, q_{12}]$ denotes the q_{12}-dimensional Wishart distribution, ρ_{01} and the positive definite matrix $\mathbf{R}_{01}$ are given hyper-parameters. Moreover, the prior distribution of $\psi_{1\delta k}$ and $\mathbf{\Lambda}_{1\omega k}$ are

$$p(\psi_{1\delta k}) \stackrel{D}{=} \text{Gamma}\,[\alpha_{01\delta k}, \beta_{01\delta k}] \ \text{ and } \ p(\mathbf{\Lambda}_{1\omega k}|\psi_{1\delta k}) \stackrel{D}{=} N[\mathbf{\Lambda}_{01\omega k}, \psi_{1\delta k}\mathbf{H}_{01\omega k}], \tag{39}$$

where $\alpha_{01\delta k}$, $\beta_{01\delta k}$, $\mathbf{\Lambda}_{01\omega k}$, and $\mathbf{H}_{01\omega k}$ are given hyper-parameters. Let $\mathbf{E}_1 = \{(\boldsymbol{\eta}_{1g1}, \ldots, \boldsymbol{\eta}_{1gN_g});\ g = 1, \ldots, G\}$, $\mathbf{E}_{1k}^T$ be the kth row of $\mathbf{E}_1$, $\Xi_1 = \{(\boldsymbol{\xi}_{1g1}, \ldots, \boldsymbol{\xi}_{1gN_g});\ g = 1, \ldots, G\}$ and $\mathbf{F}_1^* = \{(\mathbf{F}_1^*(\boldsymbol{\xi}_{1g1}), \ldots, \mathbf{F}_1^*(\boldsymbol{\xi}_{1gN_g}));\ g = 1, \ldots, G\}$, in which $F_1^*(\boldsymbol{\xi}_{1gi}) = (\boldsymbol{\eta}_{1gi}^T, \mathbf{F}_1(\boldsymbol{\xi}_{1gi})^T)^T$, $i = 1, \ldots, N_g$, and it can be shown that for $k = 1, \ldots, q_{11}$,

$$p(\psi_{1\delta k}|\cdot) \stackrel{D}{=} \text{Gamma}\,[2^{-1}N_g + \alpha_{01\delta k}, \beta_{1\delta k}],\ \ p(\mathbf{\Lambda}_{1\omega k}|\psi_{1\delta k}, \cdot) \stackrel{D}{=} N[\mathbf{m}_{1\omega k}, \psi_{1\delta k}\Sigma_{1\omega k}], \tag{40}$$

where $\Sigma_{1\omega k} = (\mathbf{H}_{01\omega k}^{-1} + \mathbf{F}_1^*\mathbf{F}_1^{*^T})^{-1}$, $\mathbf{m}_{1\omega k} = \Sigma_{1\omega k}(\mathbf{H}_{01\omega k}^{-1}\mathbf{\Lambda}_{01\omega k} + \mathbf{F}_1^*\mathbf{E}_{1k})$, and $\beta_{1\delta k} = \beta_{01\delta k} + \frac{1}{2}(\mathbf{E}_{1k}^T\mathbf{E}_{1k} - \mathbf{m}_{1\omega k}^T\Sigma_{1\omega k}^{-1}\mathbf{m}_{1\omega k} + \mathbf{\Lambda}_{01\omega k}^T\mathbf{H}_{01\omega k}\mathbf{\Lambda}_{01\omega k})$. Let $IW(\cdot, \cdot, \cdot)$ be the inverted Wishart distribution, the conditional distribution relating to $\mathbf{\Phi}_1$ is given by

$$p(\mathbf{\Phi}_1|\Xi_1) \stackrel{D}{=} IW\left[(\Xi_1\Xi_1^T + \mathbf{R}_{01}^{-1}), \sum_{g=1}^{G} N_g + \rho_{01}, q_{12}\right]. \tag{41}$$

Conditional distributions involved in $\boldsymbol{\theta}_2$ are derived similarly on the basis of the following independent conjugate type prior distributions: for $k = 1, \ldots, p$, and

$$p(\psi_{2\epsilon k}^{-1}) \stackrel{D}{=} \text{Gamma}[\alpha_{02\epsilon k}, \beta_{02\epsilon k}],\ \ p(\mathbf{\Lambda}_{2k}|\psi_{2k}) \stackrel{D}{=} N[\mathbf{\Lambda}_{02k}, \psi_{2k}\mathbf{H}_{02yk}],$$
$$p(\mathbf{A}_{2k}) \stackrel{D}{=} N[\mathbf{A}_{02k}, \mathbf{H}_{02k}],\ \ k = 1, \ldots, p,$$

where $\mathbf{A}_{2k}^T$ and $\mathbf{\Lambda}_{2k}^T$ are the vectors that contain unknown parameters in the kth rows of $\mathbf{A}_2$ and $\mathbf{\Lambda}_2$, respectively; $\alpha_{02\epsilon k}$, $\beta_{02\epsilon k}$, $\mathbf{A}_{02k}$, $\mathbf{\Lambda}_{02k}$, $\mathbf{H}_{02k}$, and $\mathbf{H}_{02yk}$ are given hyperparameters.

Similarly, conditional distributions involved in $\boldsymbol{\theta}_{2\omega}$ are derived on the basis of the following conjugate type distributions: for $k = 1, \ldots, q_{21}$,

$$p(\psi_{2\delta k}^{-1}) \stackrel{D}{=} \text{Gamma}[\alpha_{02\delta k}, \beta_{02\delta k}],\ \ \ p(\mathbf{\Lambda}_{2\omega k}|\psi_{2\delta k}) \stackrel{D}{=} N[\mathbf{\Lambda}_{02\omega k}, \psi_{2\delta k}\mathbf{H}_{02\omega k}],$$
$$p(\mathbf{\Phi}_2^{-1}) \stackrel{D}{=} W[\mathbf{R}_{02}, \rho_{02}, q_{22}],$$

where $\mathbf{\Lambda}_{2\omega} = (\mathbf{B}_2^T, \boldsymbol{\Pi}_2^T, \mathbf{\Gamma}_2^T)^T$ and $\mathbf{\Lambda}_{2\omega k}$ is the vector that contains the unknown parameters in the kth row of $\mathbf{\Lambda}_{2\omega}$. As these conditional distributions are similar to those in (38), (40) and (41), they are not presented to save space.

References

Ansari, A. and Jedidi, K. (2000). Bayesian factor analysis for multilevel binary observations. *Psychometrika*, **65**, 475–498

Austin, P. C. and Escobar, M. D. (2005). Bayesian modeling of missing data in clinical research. *Computational Statistics and Data Analysis*, **48**, 821–836

Bentler, P. M. and Bonett, D. G. (1980). Significance tests and goodness of fit in the analysis of covariance structures. *Psychological Bulletin*, **88**, 588–606

Bentler, P. M. and Stein, J. A. (1992). Structural equation models in medical research. *Statistical Methods in Medical Research*, **1**, 159–181

Bentler, P. M. and Wu, E. J. C. (2002). *EQS6 for Windows User Guide*. Encino, CA: Multivariate Software, Inc.

Bollen, K. A. (1989). *Structural Equations with Latent Variables*. New York: Wiley

Chib, S. (1995). Marginal likelihood from the Gibbs output. *Journal of the American Statistical Association*, **90**, 1313–1321

Chib, S. and Jeliazkov, I. (2001). Marginal likelihood from the Metropolis–Hastings outputs. *Journal of the American Statistical Association*, **96**, 270–281

Congdon, P. (2005). Bayesian predictive model comparison via parallel sampling. *Computational Statistics and Data Analysis*, **48**, 735–753

Diciccio, T. J., Kass, R. E., Raftery, A. and Wasserman, L. (1997). Computing Bayes factors by combining simulation and asymptotic approximations. *Journal of the American Statistical Association*, **92**, 903–915

Dolan, C., van der Sluis, S. and Grasman, R. (2005). A note on normal theory power calculation in SEM with data missing completely at random. *Structural Equation Modeling*, **12**, 245–262

Dunson, D. B. (2000). Bayesian latent variable models for clustered mixed outcomes. *Journal of the Royal Statistical Society, Series B*, **62**, 355–366

Dunson, D. B. (2005). Bayesian semiparametric isotonic regression for count data. *Journal of the American Statistical Association*, **100**, 618–627

Dunson, D. B. and Herring, A. H. (2005). Bayesian latent variable models for mixed discrete outcomes. *Biostatistics*, **6**, 11–25

Garcia-Donato, G. and Chan, M. H. (2005). Calibrating Bayes factor under prior predictive distributions. *Statistica Sinica*, **15**, 359–380

Gelfand, A. E. and Dey, D. K. (1994). Bayesian model choice: asymptotic and exact calculations. *Journal of the Royal Statistical Society, Series B*, **56**, 501–514

Gelman, A. and Meng, X. L. (1998). Simulating normalizing constants: from importance sampling to bridge sampling to path sampling. *Statistical Science*, **13**, 163–185

Gelman, A., Meng, X. L. and Stern, H. (1996). Posterior predictive assessment of model fitness via realized discrepancies. *Statistica Sinica*, **6**, 733–807

Geman, S. and Geman, D. (1984) Stochastic relaxation, Gibbs distributions, and the Bayesian restoration of images. *IEEE Transactions on Pattern Analysis and Machine Intelligence*, **6**, 721–741

Hastings, W. K. (1970). Monte Carlo sampling methods using Markov chains and their application. *Biometrika*, **57**, 97–109

Jöreskog, K. G. and Sörbom, D. (1996). *LISREL 8: Structural Equation Modeling with the SIMPLIS Command Language*. Scientific Software International: Hove and London

Kass, R. E. and Raftery, A. E. (1995). Bayes factors. *Journal of the American Statistical Association*, **90**, 773–795

Kim, K. H. (2005). The relation among fit indices, power, and sample size in structural equation modeling. *Structural Equation Modeling*, **12**, 368–390

Lee, S. Y. (2007). *Structural Equations Modelling: A Bayesian Approach*. New York: Wiley

Lee, S. Y. and Song, X. Y. (2003). Model comparison of a nonlinear structural equation model with fixed covariates. *Psychometrika*, **68**, 27–47

Lee, S. Y. and Song, X. Y. (2004a). Evaluation of the Bayesian and maximum likelihood approaches in analyzing structural equation models with small sample sizes. *Multivariate Behavioral Research*, **39**, 653–686

Lee, S. Y. and Song, X. Y. (2004b). Maximum likelihood analysis of a general latent variable model with hierarchically mixed data. *Biometrics*, **60**, 624–636

Lee, S. Y. and Xia, Y. M. (2006). Maximum likelihood methods in treating outliers and symmetrically heavy-tailed distributions for nonlinear structural equation models with missing data. *Psychometrika*, **71**, 565–585

Levin, H. M. (1998). Accelerated schools: a decade of evolution. In A. Hargreavers et al. (Eds) *International Handbook of Educational Change, Part Two* (pp 809–810). New York: Kluwer

Liu, X., Wall, M. M. and Hodges, J. S. (2005). Generalized spatial structural equation models. *Biostatistics*, **6**, 539–551

Meng, X. L. and Wong, H. W. (1996). Simulating ratios of normalizing constants via a simple identity: a theoretical exploration. *Statistica Sinica*, **6**, 831–860

Metropolis, N., Rosenbluth, A. W., Rosenbluth, M. N., Teller, A. H. and Teller, E. (1953). Equations of state calculations by fast computing machine. *Journal of Chemical Physics*, **21**, 1087–1091

Ogata, Y. (1989). A Monte Carlo method for high dimensional integration. *Numerische Mathematik*, **55**, 137–157

Palomo, J., Dunson, D. B. and Bollen, K. (2007). Bayesian structural equation modeling. In S. Y. Lee (Ed) *Handbook of Latent Variable and Related Models*. Amsterdam: Elsevier

Pugesek, B. H., Tomer, A. and von Eye, A. (2003). *Structural Equation Modeling Applications in Ecological and Evolutionary Biology*. New York: Cambridge University Press

Raftery, A. E. (1993). Bayesian model selection in structural equation models. In K. A. Bollen and J. S. Long (Eds) *Testing Structural Equation Models* (pp 163–180). Thousand Oaks, CA: Sage Publications

Raftery, A. E. (1996). Hypothesis testing and model selection. In W. R. Wilks, S. Richardson and D. J. Spieglhalter (Eds) *Practical Markov Chain Monte Carlo* (pp 163–188). London: Chapman and Hall

Raykov, T. and Marcoulides, G. A. (2006). On multilevel model reliability estimation from the perspective of structural equation modeling. *Structural Equation Modeling*, **13**, 130–141

Richardson, S. and Green, P. J. (1997). On Bayesian analysis of mixture with an unknown number of components (with discussion). *Journal of the Royal Statistical Society, Series B*, **59**, 731–792

Sanchez, B. N., Budtz-Jorgenger, E., Ryan, L. M. and Hu, H. (2005). Structural equation models: a review with applications to environmental epidemiology. *Journal of the American Statistical Association*, **100**, 1443–1455

Schwarz, G. (1978). Estimating the dimension of a model. *The Annals of Statistics*, **6**, 461–464

Shi, J. Q. and Lee, S. Y. (2000). Latent variable models with mixed continuous and polytomous data. *Journal of the Royal Statistical Society, Series B*, **62**, 77–87

Song, X. Y. and Lee, S. Y. (2004). Bayesian analysis of two-level nonlinear structural equation models with continuous and polytomous data. *British Journal of Mathematical and Statistical Psychology*, **57**, 29–52

Song, X. Y. and Lee, S. Y. (2005). A multivariate probit latent variable model for analyzing dichotomous responses. *Statistica Sinica*, **15**, 645–664

Song, X. Y. and Lee, S. Y. (2006a). Model comparison of generalized linear mixed models. *Statistics in Medicine*, **25**, 1685–1698

Song, X. Y. and Lee, S. Y. (2006b). Bayesian analysis of latent variable models with non-ignorable missing outcomes from exponential family. *Statistics in Medicine*, **26**, 681–693

Spiegelhalter, D. J., Best, N. G., Carlin, B. P. and van der Linde, A. (2002). Bayesian measures of model complexity and fit. *Journal of the Royal Statistical Society, Series B*, **64**, 583–639

Spiegelhalter, D. J., Thomas, A., Best, N. G. and Lunn, D. (2003). *WinBugs User Manual. Version 1.4.* Cambridge, England: MRC Biostatistics Unit

Tanner, M. A. and Wong, W. H. (1987). The calculation of posterior distributions by data augmentation (with discussion). *Journal of the American statistical Association*, **82**, 528–550

Bayesian Model Selection in Factor Analytic Models

Joyee Ghosh and David B. Dunson

1 Introduction

Factor analytic models are widely used in social science applications to study latent traits, such as intelligence, creativity, stress, and depression, that cannot be accurately measured with a single variable. In recent years, there has been a rise in the popularity of factor models due to their flexibility in characterizing multivariate data. For example, latent factor regression models have been used as a dimensionality reduction tool for modeling of sparse covariance structures in genomic applications (West, 2003; Carvalho et al., 2008). In addition, structural equation models and other generalizations of factor analysis are widely useful in epidemiologic studies involving complex health outcomes and exposures (Sanchez et al., 2005). Improvements in Bayesian computation permit the routine implementation of latent factor models via Markov chain Monte Carlo (MCMC) algorithms, and a very broad class of models can be fitted easily using the freely available software package WinBUGS. The literature on methods for fitting and inferences in latent factor models is vast (for recent books, see Loehlin, 2004; Thompson, 2004).

In using a factor analytic model for inferences on a covariance structure, it is appealing to formally account for uncertainty in selecting the number of factors. There has been some focus in the frequentist and Bayesian literature on the problem of selection of the number of factors. Press and Shigemasu (1999) propose to choose the number of factors having the highest posterior probability, noting that such an approach improves upon the commonly used AIC (Akaike, 1987) and

J. Ghosh
Department of Statistical Science, Duke University, Durham, NC 27708-0251
joyee@stat.duke.edu

D.B. Dunson
Biostatistics Branch, National Institute of Environmental Health Sciences, RTP, NC 27709
and
Department of Statistical Science, Duke University, Durham, NC 27708-0251
dunson1@niehs.nih.gov

D. B. Dunson (ed.) *Random Effect and Latent Variable Model Selection*,
DOI: 10.1007/978-0-387-76721-5,

BIC (Schwarz, 1978) criteria. For hierarchical models, such as latent factor models, the BIC justification as an approximation to the Bayes factor breaks down (Berger et al., 2003), and one may need a different penalty for model complexity (Zhang and Kocka, 2004).

Estimation of posterior probabilities of models having different numbers of latent factors poses major challenges such as (a) how to choose priors for the factor loadings in the list of models corresponding to different numbers of factors; and (b) how to efficiently and accurately estimate posterior model probabilities. Polasek (1997) considered approaches for estimating posterior probabilities based on separate MCMC analyses of models differing only in the number of factors. Although an estimate of the marginal likelihood is not automatically available from the MCMC output, a number of algorithms have been proposed (Chib, 1995; DiCiccio et al., 1997; Gelfand and Dey, 1994; Meng and Wong, 1996).

Lopes and West (2004) proposed a reversible jump MCMC (RJMCMC) algorithm (Green, 1995) to move between the models with different numbers of factors, and conducted a thorough comparison with estimators for approximating marginal likelihoods from separate MCMC analyses under each model. In simulation studies, they found that a number of the methods perform poorly relative to the RJMCMC and bridge sampling (Meng and Wong, 1996) in terms of proportions of simulations in which the true number of factors is assigned highest posterior probability. A computational challenge in implementing RJMCMC for factor model selection is the difficulty of choosing efficient proposal distributions. Lopes and West (2004) address this problem by constructing proposals using the results of a preliminary MCMC run under each model. Such an approach is highly computationally demanding, becoming infeasible as the sample size and potential number of factors increases. Motivated by this problem, Carvalho et al. (2008) proposed an evolutionary search algorithm, which provides a useful approach for searching for good factor models in high dimensions.

Lee and Song (2002) developed a method for estimating Bayes factors for selecting the number of factors in a factor analysis model using the idea of path sampling (Gelman and Meng, 1998). They proposed a procedure that is simple to implement and tends to have good performance in terms of accuracy. To estimate a single Bayes factor for comparing two competing models, their method requires running of separate MCMC algorithms along a grid corresponding to different values for a path sampling constant. Although such an approach can potentially be implemented in parallel, computational efficiency is nonetheless a concern, particularly if one has many different models under consideration. In addition, if the individual Markov chains exhibit poor mixing, which is a common problem in latent factor models, one may need to run each chain for a very large number of iterations to obtain accurate results.

An additional issue is that it is well known that Bayes factors are sensitive to the choice of prior. In latent factor models, it tends to be difficult to elicit the parameters, since in typical applications there is substantial uncertainty about the true values of the factor loadings and error variances a priori. Hence, one would often prefer to choose a vague prior. However, vague priors lead to problems with MCMC convergence, since in the limiting case as the prior variance becomes large, one can obtain an improper posterior. In addition, even if this was not an issue, high variance priors tend to systematically favor small models, so that one will tend to select a one-factor model if the prior variance is extremely large. In implementing their approach, Lee and Song (2002) used highly informative priors to avoid this problem. However, if their highly informative values are concentrated around the wrong values, as would typically be the case if there is substantial uncertainty a priori, then one would expect poor performance in terms of model selection. For example, one may tend to discard an important factor inappropriately if the priors for the factor loadings have small variance around values far from the truth.

Ghosh and Dunson (2007) proposed an approach for simultaneously addressing the issues of efficient computation and prior specification in latent factor models through the use of a parameter-expansion approach. This approach conveys a dramatic improvement in MCMC efficiency in many cases, while inducing heavy-tailed priors that can be used for robust model selection via the path sampling approach. We have observed good performance of such an approach in a number of simulation studies.

In this chapter, we review the Ghosh and Dunson (2007) approach and propose a new approach for estimating posterior probabilities of models with different numbers of factors. The proposed approach relies on development of parameter expanded Gibbs samplers (Liu and Wu, 1999; Gelman et al., 2007; Ghosh and Dunson, 2007) for all models in our list. Based on the MCMC output, we compute estimates for Bayes factors for models differing by one factor. Using a simple identity, one can then compute posterior probabilities from these estimates. Clearly, this method is computationally less demanding than some of the other methods as we need to run a single MCMC algorithm to estimate a particular Bayes factor. Additionally, the use of parameter expansion facilitates good mixing. The method provides reasonably accurate results, based on simulation studies.

Section 2 defines the model when the number of factors is known. Section 3 describes the methodology used for estimating Bayes factors based on MCMC output when the number of factors is unknown. Section 4 presents the results of a simulation study. Section 5 contains an application to rodent organ weight data, and Sect. 6 discusses the results. The full conditional distributions for the parameter expanded Gibbs Sampler are given in the Appendix.

2 Specification of the Model

We shall first define a factor model in which the number of factors is known to be k,

$$\mathbf{y}_i = \boldsymbol{\Lambda}\boldsymbol{\eta}_i + \boldsymbol{\epsilon}_i, \quad \boldsymbol{\epsilon}_i \sim \mathrm{N}_p(\mathbf{0}, \boldsymbol{\Sigma}), \tag{1}$$

where $\boldsymbol{\Lambda}$ is a $p \times k$ matrix of factor loadings, $\boldsymbol{\eta}_i = (\eta_{i1}, \ldots, \eta_{ik})' \sim \mathrm{N}_k(\mathbf{0}, \mathbf{I}_k)$ a vector of standard normal latent factors, and $\boldsymbol{\epsilon}_i$ is a residual with diagonal covariance matrix $\boldsymbol{\Sigma} = \text{diag}(\sigma_1^2, \ldots, \sigma_p^2)$. Here the underlying latent factors, $\boldsymbol{\eta}_i$, induce dependence among the components of $\mathbf{y}_i$, because integrating out the latent factors we can write the marginal distribution of $\mathbf{y}_i$ as $\text{N}_p(\mathbf{0}, \boldsymbol{\Omega})$, with $\boldsymbol{\Omega} = \boldsymbol{\Lambda}\boldsymbol{\Lambda}' + \boldsymbol{\Sigma}$. Thus, this model implies that the sharing of common latent factors explains the dependence in the outcomes, and given the latent factors, the outcome variables are uncorrelated. For example, the outcomes can be the results of various blood tests of an individual and the underlying latent factor may be the health score for that individual. Usually the number of factors is small, relative to the number of outcomes ($k \ll p$). This leads to sparse models for $\boldsymbol{\Omega}$ containing fewer than $p(p+1)/2$ parameters. For this reason, factor models provide a convenient and flexible framework for modeling a covariance matrix, especially in applications with moderate to large p.

The above factor model (1) without further constraints is not identifiable under orthogonal rotation. For example, if we post-multiply $\boldsymbol{\Lambda}$ by an orthonormal matrix $\mathbf{P}$, where $\mathbf{P}$ is such that $\mathbf{P}\mathbf{P}' = \mathbf{I}_k$, we will obtain exactly the same $\boldsymbol{\Omega}$ as in the previous factor model (1). To avoid this issue of non-identifiability, we impose some additional restrictions on the factor loadings matrix $\boldsymbol{\Lambda}$. We assume that $\boldsymbol{\Lambda}$ has a full-rank lower triangular structure to ensure identifiability. As the $k(k-1)/2$ elements in the upper triangular part of $\boldsymbol{\Lambda}$ are restricted to be zero, the number of free parameters in $\boldsymbol{\Lambda}$ and $\boldsymbol{\Sigma}$ is $q = p(k+1) - k(k-1)/2$. Here, k must be chosen so that $q \leq p(p+1)/2$. We use this restriction as a default, motivated by applications in which the latent factors do not have a pre-specified interpretation but are included primarily to induce a sparse covariance structure in multivariate data. However, our approach can be easily modified to factor analysis applications in which the different measurements are designed as manifestations of latent traits of interest (e.g., intelligence, stress, etc.).

To complete a Bayesian specification of model (1) we would need to specify the prior distributions for the free elements of $\boldsymbol{\Lambda}$ and Σ. A popular choice is truncated normal priors for the diagonal elements of $\boldsymbol{\Lambda}$, normal priors for the lower triangular elements, and inverse-gamma priors for $\sigma_1^2, \ldots, \sigma_p^2$. These choices are convenient, because they represent conditionally conjugate forms that lead to straightforward posterior computation by a Gibbs sampler (Arminger, 1998; Rowe, 1998; Song and Lee, 2001). However, this prior specification suffers from a few major drawbacks. First, specification of the hyperparameters in the prior may be difficult. Prior elicitation is particularly important in this model, because in the limiting case as the prior variance for the normal and inverse-gamma components increases the posterior becomes improper. To address this problem, often informative priors are chosen. In the absence of subject-matter knowledge, sometimes the hyperparameters in the prior are chosen after an initial analysis of the data. Using the data twice in this manner could lead to an underestimation of uncertainty in model selection. Second, even if informative priors are used, the Gibbs samplers tend to exhibit extreme slow-mixing.

Ghosh and Dunson (2007) address the above problems by generalizing the idea of Gelman (2006) to induce a new class of robust priors for the factor loadings. They use parameter expansion to induce a class of t or folded-t priors depending on sign constraints on the loadings. In absence of subject-matter expertise, they recommend using a Cauchy or half-Cauchy prior as a default. They also demonstrate good mixing properties of their parameter expanded Gibbs sampler compared to the traditional Gibbs sampler, using various simulated and real data examples. In this chapter, we outline a method to select the number of factors using their prior.

3 Bayesian Uncertainty in the Number of Factors

To allow an unknown number of factors k, we choose a multinomial prior distribution, with $\Pr(k = h) = \kappa_h$, for $h = 1, \ldots, m$. We then complete a Bayesian specification through priors on the coefficients within each of the models in the list $k \in \{1, \ldots, m\}$. This is accomplished by choosing a prior for the coefficients in the m factor model having the form described in Ghosh and Dunson (2007). We first induce a prior on $\Lambda^{(m)}$ through parameter expansion. For any smaller model $k = h$, the prior for $\Lambda^{(h)}$ is obtained by marginalizing out the columns from $(h + 1)$ to m. In this manner, we place a prior on the coefficients in the largest model, while inducing priors on the coefficients in each of the smaller models. We induce a prior on $\Lambda^{(m)}$ by defining the following parameter expanded (PX) factor model:

$$\mathbf{y}_i = \mathbf{\Lambda}^{*(m)}\boldsymbol{\eta}_i^* + \boldsymbol{\epsilon}_i, \quad \boldsymbol{\eta}_i^* \sim \mathrm{N}_k(\mathbf{0}, \mathbf{\Psi}), \quad \boldsymbol{\epsilon}_i \sim \mathrm{N}_p(\mathbf{0}, \mathbf{\Sigma}), \tag{2}$$

where $\mathbf{\Lambda}^{*(m)}$ is $p \times m$ working factor loadings matrix having a lower triangular structure without constraints on the elements, $\boldsymbol{\eta}_i^* = (\eta_{i1}^*, \ldots, \eta_{im}^*)'$ is a vector of working latent variables, $\mathbf{\Psi} = \mathrm{diag}(\psi_1, \ldots, \psi_m)$, and $\mathbf{\Sigma}$ is a diagonal covariance matrix defined as in (1). Note that model (2) is clearly over-parameterized having redundant parameters in the covariance structure. In particular, marginalizing out the latent variables, $\boldsymbol{\eta}_i^*$, we obtain $\mathbf{y}_i \sim \mathrm{N}_p(\mathbf{0}, \mathbf{\Lambda}^{*(m)}\mathbf{\Psi}\mathbf{\Lambda}^{*(m)'} + \mathbf{\Sigma})$. Clearly, the diagonal elements of $\mathbf{\Lambda}^{*(m)}$ and $\mathbf{\Psi}$ are redundant.

To relate the working model parameters in (2) to the inferential model parameters in (1), we use the following transformation:

$$\lambda_{jl}^{(m)} = \mathcal{S}(\lambda_{ll}^{*(m)})\lambda_{jl}^{*(m)}\psi_l^{1/2}, \quad \eta_{il} = \mathcal{S}(\lambda_{ll}^{*(m)})\psi_l^{-1/2}\eta_{il}^* \quad \text{for} \quad j = 1, \ldots, p,$$
$$l = 1, \ldots, m, \tag{3}$$

where $\mathcal{S}(x) = -1$ for $x < 0$ and $\mathcal{S}(x) = 1$ for $x \geq 0$. Then, instead of specifying a prior for $\mathbf{\Lambda}^{(m)}$ directly, we induce a prior on $\mathbf{\Lambda}^{(m)}$ through a prior for $\mathbf{\Lambda}^{*(m)}$, $\mathbf{\Psi}$. In particular, we let

$$\lambda_{jl}^{*(m)} \overset{iid}{\sim} \mathrm{N}(0, 1), \; j = 1, \ldots, p, \; l = 1, \ldots, \min(j, m),$$

$$\lambda_{jl}^{(m)} \sim \delta_0, \; j = 1, \ldots, (m-1), \; l = j+1, \ldots, m, \; \psi_l \overset{iid}{\sim} \mathcal{G}(a_l, b_l), \; l = 1, \ldots, m, \tag{4}$$

where δ_0 is a measure concentrated at 0, and $\mathcal{G}(a, b)$ denotes the gamma distribution with mean a/b and variance a/b^2. This prior is conditionally conjugate, leading to straightforward Gibbs sampling.

Bayesian selection of the number of factors relies on posterior model probabilities

$$\Pr(k = h \mid \mathbf{y}) = \frac{\kappa_h \, \pi(\mathbf{y} \mid k = h)}{\sum_{l=1}^{m} \kappa_l \, \pi(\mathbf{y} \mid k = l)}, \tag{5}$$

where the marginal likelihood under model k, $\pi(\mathbf{y} \mid k = h)$, is obtained by integrating the likelihood $\prod_i \mathrm{N}_p(\mathbf{y}_i; \mathbf{0}, \mathbf{\Lambda}^{(k)}\mathbf{\Lambda}^{(k)'} + \mathbf{\Sigma})$ across the prior for the factor loadings $\mathbf{\Lambda}^{(k)}$ and residual variances $\mathbf{\Sigma}$. We still need to consider the problem of estimating $\Pr(k = h \mid \mathbf{y})$ as the marginal likelihood is not available in closed form. Note that any posterior model probability can be expressed entirely in terms of the prior odds $O[h : j] = \{\kappa_h/\kappa_j\}$ and Bayes factors $BF[h : j] = \{\pi(\mathbf{y} \mid k = h)/\pi(\mathbf{y} \mid k = j)\}$ as follows:

$$\Pr(k = h \mid \mathbf{y}) = \frac{O[h : j] * BF[h : j]}{\sum_{l=1}^{m} O[l : j] * BF[l : j]}. \tag{6}$$

We choose $\kappa_h = 1/m$, which corresponds to a uniform prior for the number of factors k. To obtain an estimate of the Bayes factor BF$[(h-1) : h]$, for comparing models $k = (h-1)$ to $k = h$, we run the PX Gibbs sampler under $k = h$. Let $\{\boldsymbol{\theta}_i^{(h)}, i = 1, \ldots, I\}$ denote the I MCMC samples from the PX Gibbs sampler under $k = h$, where $\boldsymbol{\theta}_i^{(h)} = (\mathbf{\Lambda}_i^{(h)}, \mathbf{\Sigma}_i)$. We can then estimate $BF[(h-1) : h]$ by the following estimator:

$$\widehat{BF}[(h-1) : h] = \frac{1}{I} \sum_{i=1}^{I} \frac{p(\mathbf{y}|\boldsymbol{\theta}_i^{(h)}, k = h-1)}{p(\mathbf{y}|\boldsymbol{\theta}_i^{(h)}, k = h)}. \tag{7}$$

We do this for $h = 2, \ldots, m$.

The above estimator is based on the following identity:

$$\begin{aligned} &\int \frac{p(\mathbf{y}|\boldsymbol{\theta}^{(h)}, k = h-1)}{p(\mathbf{y}|\boldsymbol{\theta}^{(h)}, k = h)} p(\boldsymbol{\theta}^{(h)}|\mathbf{y}, k = h) d\boldsymbol{\theta}^{(h)} \\ &\quad = \int p(\mathbf{y}|\boldsymbol{\theta}^{(h)}, k = h-1) \frac{p(\boldsymbol{\theta}^{(h)})}{p(\mathbf{y}|k = h)} d\boldsymbol{\theta}^{(h)}, \\ &\quad = \frac{p(\mathbf{y}|k = h-1)}{p(\mathbf{y}|k = h)} \end{aligned} \tag{8}$$

We can obtain the Bayes factor for comparing any two models. For example, the Bayes factor for comparing the one-factor and the m-factor models is obtained as: $BF[1 : m] = BF[1 : 2] * BF[2 : 3] \ldots BF[(m-1) : m]$. Using (6), we can estimate the posterior model probabilities in (5). We will refer to this approach as importance sampling with parameter expansion (IS-PX).

Lee and Song (2002) use the path sampling approach of Gelman and Meng (1998) for estimating log Bayes factors. They construct a path using a scalar $t \in [0, 1]$ to link two models M_0 and M_1. They use the same idea as outlined in an example in Gelman and Meng (1998) to construct their path. To compute the required integral, they take a fixed set of grid points for t, $t \in [0, 1]$ and then use numerical integration to approximate the integration over t. For a detailed description of using path sampling for computing Bayes factors refer to the chapter *Bayesian Model Comparison of Structural Equation Models* in this book. Note that factor models are a special case of structural equation models.

Although their approach is promising in terms of accuracy, it is quite computationally intensive, requiring running of separate MCMC algorithms for each value of t in the grid. In contrast, using IS-PX we avoid the need to run multiple analyses, though it is not clear that this will necessarily improve efficiency, since we may require a long chain to obtain accurate estimates of the Bayes factors. Given that the true Bayes factors are not available and are analytically intractable, our assessment of the performance of IS-PX will rely on simulations.

In addition to the difficulty of estimating the Bayes factor, an important challenge in Bayes model comparisons is sensitivity to the prior. It is well known that Bayes factors tend to be sensitive to the prior, motivating a rich literature on objective Bayes methods (Berger and Pericchi, 1996, 2001). Lee and Song (2002) rely on highly informative priors in implementing Bayesian model selection for factor analysis, an approach which is only reliable when substantial prior knowledge is available allowing one to concisely guess a narrow range of plausible values for all of the parameters in the model. Such knowledge is often lacking. This motivated Ghosh and Dunson (2007) to modify the Lee and Song (2002) path sampling approach to allow the use of their default PX-induced priors. They refer to this as path sampling with parameter expansion (PS-PX).

4 Simulation Study

Here we consider two sets of simulation studies and compare our results with those from the PS-PX approach, as reported in Ghosh and Dunson (2007). Let m denote the maximum number of factors in our list. We routinely standardize the data prior to analysis.

4.1 One-Factor Model

In the first simulation, $p = 7$, $n = 100$, the true number of factors, $k = 1$ and

$$\begin{aligned}\boldsymbol{\Lambda} &= (0.995, 0.975, 0.949, 0.922, 0.894, 0.866, 0.837)',\\ \text{diag}(\boldsymbol{\Sigma}) &= (0.01, 0.05, 0.10, 0.15, 0.20, 0.25, 0.30).\end{aligned}$$

we take m to be 3, which is also the maximum number of factors resulting in an identifiable model. We repeat this simulation for 100 simulated data sets. To specify the prior for the PX Gibbs sampler, we induce half-Cauchy and Cauchy priors, both with a scale parameter 1, for the diagonals and lower triangular elements of $\boldsymbol{\Lambda}$, respectively. For the residual precisions σ_j^{-2} we take $\mathcal{G}(1, 0.2)$ priors. This choice of hyperparameter values provides a modest degree of shrinkage towards a plausible range of values for the residual precision. For each simulated data set, we run the Gibbs sampler for 25,000 iterations, discarding the first 5,000 iterations as a burn-in. We note that for IS-PX we need only the samples from the grid $t = 1$ of the PS-PX approach. Here IS-PX chooses the correct model 92/100 times and PS-PX 100/100 times.

4.2 Three-Factor Model

For the second simulation study, $p = 10$, $n = 100$, and the true number of factors, $k = 3$. This presents a more difficult scenario as some of the loadings are negative and there is more noise in the data compared to the previous one-factor model.

$$\boldsymbol{\Lambda}' = \begin{pmatrix} 0.89 & 0.00 & 0.25 & 0.00 & 0.80 & 0.00 & 0.50 & 0.00 & 0.00 & 0.00 \\ 0.00 & 0.90 & 0.25 & 0.40 & 0.00 & 0.50 & 0.00 & 0.00 & -0.30 & -0.30 \\ 0.00 & 0.00 & 0.85 & 0.80 & 0.00 & 0.75 & 0.75 & 0.00 & 0.80 & 0.80 \end{pmatrix},$$

$$\text{diag}(\boldsymbol{\Sigma}) = (0.2079, 0.1900, 0.1525, 0.2000, 0.3600, 0.1875, 0.1875, 1.0000, 0.2700, 0.2700).$$

For this second simulation example the true model has three factors and the maximum number of factors resulting in an identifiable model is 6. We take $m = 4$, following Ghosh and Dunson (2007). We carry out the simulations exactly as in the previous simulation study. Here IS-PX chooses the correct model 86/100 times compared to 100/100 by PS-PX. In these simulations ten gridpoints were considered for PS-PX, so given that IS-PX takes only 1/10 of the run-time its performance is reasonably good. Running the MCMC much longer would improve its performance. A point to be noted here is that in the simulations where the wrong model is chosen, the bigger model with four factors is selected. For these datasets the parameter estimates under both three- and four-factor models are very similar, with the entries in the fourth column of the loadings matrix being close to zero. Hence even if the bigger model is chosen by IS-PX in some cases, we do not expect much deterioration in parameter estimates.

5 Application to Rodent Organ Weight Data

We illustrate our method for model selection by using it on a real dataset. We have organ weight data from a US National Toxicology Program (NTP) 13-week study of Anthraquinone in female Fischer rats. The goal of such studies is to assess the short-term toxicological effects of test agents on a variety of outcomes, including animal and organ body weights. Studies are routinely conducted with 60 animals randomized to approximately six dose groups, including a control. In the Anthraquinone study, doses included 0, 1,875, 3,750, 7,500, 15,000 and 30,000 ppm.

At the end of the study, animals are sacrificed and a necropsy is conducted, with overall body weight obtained along with weights for the heart, liver, lungs, kidneys (combined), and thymus. Although body and organ weights are clearly correlated, a challenge in the analysis of these data is the dimensionality of the covariance matrix. In particular, even assuming a constant covariance across dose groups, it is still necessary to estimate $p(p+1)/2 = 21$ covariance parameters using data from only $n = 60$ animals. Hence, routine analyses rely on univariate approaches applied separately to body weight and the different organ weights.

We can use a factor model here to reduce dimensionality, but it is not clear whether it is appropriate to assume a single factor underlying the different weights or if additional factors need to be introduced. To address this question using the Anthraquinone data, we repeated the approach described in Sect. 3 in the same manner as implemented in the simulation examples. Body weights were standardized within each dose group prior to analysis for the purposes of studying the correlation structure. Here we ran the algorithm for 100,000 iterations for IS-PX. The maximum possible number of factors was $m = 3$.

The estimated probabilities for the one-, two-, and three-factor models under IS-PX are 0.4417, 0.3464, and 0.2120 and using PS-PX are 0.9209, 0.0714, and 0.0077, respectively, as reported in Ghosh and Dunson (2007). The estimated factor loadings for the one- and two-factor models are presented in Tables 1 and 2.

Table 1 Posterior summaries under the one-factor model for organ weight data

Weight	Parameter	Mean	95% CI
Body	λ_1	0.88	[0.67,1.10]
Heart	λ_2	0.33	[0.08,0.59]
Liver	λ_3	0.52	[0.28,0.77]
Lungs	λ_4	0.33	[0.08,0.59]
Kidneys	λ_5	0.70	[0.48,0.94]
Thymus	λ_6	0.42	[0.17,0.68]

Table 2 Posterior summaries under the two-factor model for organ weight data

Weight	Parameter	Mean	95% CI	Parameter	Mean	95% CI
Body	λ_{11}	0.87	[0.66,1.09]	λ_{12}	0	[0,0]
Heart	λ_{21}	0.34	[0.08,0.61]	λ_{22}	0.24	[0.01,0.73]
Liver	λ_{31}	0.52	[0.28,0.78]	λ_{32}	−0.15	[−0.63,0.45]
Lungs	λ_{41}	0.35	[0.09,0.62]	λ_{42}	0.29	[−0.62,0.88]
Kidneys	λ_{51}	0.70	[0.48,0.94]	λ_{52}	−0.06	[−0.44,0.29]
Thymus	λ_{61}	0.42	[0.17,0.68]	λ_{62}	−0.14	[−0.66,0.49]

Body weight and kidney weight are the two outcomes having the highest correlation with the latent factor in the one-factor analysis. The estimated covariance matrix is similar under the one- and two-factor models, and the loadings on the second factor tend to be zero, suggesting that the second factor is not needed. This is also true for the third model. We also examined the likelihoods at the posterior mean of Ω. For the two- and three-factor models the increase in likelihood seems small compared to the number of additional parameters. In that sense, the factor model with one factor seems most appropriate. Hence, the estimates of the posterior probabilities from PS-PX seem more realistic. However, given that we obtained similar estimates of the induced covariance matrix under the one-, two- and three-factor models using the PX approach, it may be that there is little penalty to be paid for the additional complexity involved in fitting the multiple factor models given the regularization implicit in the Bayes approach. However, from the standpoint of ease in interpretation, the more parsimonious one-factor model is certainly preferred.

6 Discussion

For analyzing high-dimensional, or even moderate-dimensional, multivariate data, the factor model provides a convenient tool for sparse modeling of the covariance matrix. This kind of data arises in a wide variety of applications ranging from genomics, where the outcomes may be highly correlated measurements on multiple genes, to epidemiology where the goal may be to study the effect of a collection of nutrients, many of which are highly correlated. Usually in such cases, the number of factors to be included in the model is unknown and leads to a challenging model uncertainty problem. A Bayesian solution to this problem proceeds by treating the number of factors as unknown and then assigning prior probabilities to each model in the list. After observing the data, the prior probabilities are updated to obtain posterior probabilities for each model.

In this chapter, we have proposed an easy to implement, fully Bayesian approach for estimating the posterior model probabilities via Bayes factors. Model selection based on fully Bayesian approaches tends to be computationally intensive. Gibbs samplers that exhibit slow-mixing add to the already existing heavy computational burden of model selection. This is why we choose a default heavy-tailed prior for factor loadings (Ghosh and Dunson, 2007), that greatly facilitates mixing. Using a simple identity, we show our estimate is unbiased for estimating Bayes factors.

Comparing our method to the path sampling approach (Ghosh and Dunson, 2007; Song and Lee, 2002) based on simulated and real data, we find that it is relatively fast but less accurate. We have found that when it fails to choose the correct model, it usually prefers a model with more factors than the true model, and most of the loadings for the extra factors are close to zero. Since the loadings matrix for the true model can be thought to be nested within a larger factor model, with entries in the last columns equal to zero, we note that choosing a larger model in this case is more desirable than choosing a wrong model with less factors. To estimate the Bayes factor between two models that differ by one factor, our method currently uses the samples only from the larger model. Introducing a single bridge density between the two models may improve results substantially, while adding only a negligible computational cost. This still needs further work and seems to be a promising direction for future research.

Acknowledgements This research was supported by the Intramural Research Program of the NIH, National Institute of Environmental Health Sciences.

References

Akaike, H. (1987). Factor analysis and AIC. *Psychometrika* **52**, 317–332

Arminger, G. (1998). A Bayesian approach to nonlinear latent variable models using the Gibbs sampler and the Metropolis-Hastings algorithm. *Psychometrika* **63**, 271–300

Berger, J. and Pericchi, L. (1996). The intrinsic Bayes factor for model selection and prediction. *Journal of the American Statistical Association* **91**, 109–122

Berger, J. and Pericchi, L. (2001). Objective Bayesian methods for model selection: introduction and comparison [with discussion]. In: *Model Selection*, P. Lahiri (ed.). Institute of Mathematical Statistics Lecture Notes, Monograph Series Volume 38, Beachwood Ohio, 135–207

Berger, J.O., Ghosh, J.K. and Mukhopadhyay, N. (2003). Approximation and consistency of Bayes factors as model dimension grows. *Journal of Statistical Planning and Inference* **112**, 241–258

Carvalho, C., Lucas, J., Wang, Q., Nevins, J. and West, M. (2008). High-dimensional sparse factor modelling: applications in gene expression genomics. *Journal of the American Statistical Association*, to appear

Chib, S. (1995). Marginal likelihoods from the Gibbs output. *Journal of the American Statistical Association* **90**, 1313–1321

DiCiccio, T.J., Kass, R., Raftery, A. and Wasserman, L. (1997). Computing Bayes factors by combining simulations and asymptotic approximations. *Journal of the American Statistical Association* **92**, 903–915

Gelfand, A.E. and Dey, D.K. (1994). Bayesian model choice: asymptotics and exact calculations. *Journal of the Royal Statistical Society* **B**, 501–514

Gelfand, A.E., Sahu, S.K. and Carlin, B.P. (1995). Efficient parameterisations for normal linear mixed models. *Biometrika* **82**, 479–488

Gelman, A. (2006). Prior distributions for variance parameters in hierarchical models. *Bayesian Analysis* **3**, 515–534

Gelman, A. and Meng, X.L. (1998). Simulating normalizing constants: from importance sampling to bridge sampling to path sampling. *Statistical Science* **13**, 163–185

Gelman, A., van Dyk, D., Huang, Z. and Boscardin, W.J. (2007). Using redundant parameters to fit hierarchical models. *Journal of Computational and Graphical Statistics*, to appear

Ghosh, J. and Dunson, D.B. (2007). Default priors and efficient posterior computation in Bayesian factor analysis. *Journal of Computational and Graphical Statistics*, revision requested

Green, P.J. (1995). Reversible jump Markov chain Monte Carlo and Bayesian model determination. *Biometrika* **82**, 711–732

Lee, S.Y. and Song, X.Y. (2002). Bayesian selection on the number of factors in a factor analysis model. *Behaviormetrika* **29**, 23–40

Liu, J. and Wu, Y.N. (1999). Parameter expansion for data augmentation. *Journal of the American Statistical Association* **94**, 1264–1274

Loehlin, J.C. (2004). *Latent Variable Models: An Introduction to Factor, Path and Structural Equation Analysis*. Lawrence Erlbaum Associates,

Lopes, H.F. and West, M. (2004). Bayesian model assessment in factor analysis. *Statistica Sinica* **14**, 41–67
Meng, X.L. and Wong, W.H. (1996). Simulating ratios of normalising constants via a simple identity. *Statistica Sinica* **11**, 552–586
Polasek, W. (1997). Factor analysis and outliers: a Bayesian approach. Discussion Paper, University of Basel
Press, S.J. and Shigemasu, K. (1999). A note on choosing the number of factors. *Communications in Statistics – Theory and Methods* **28**, 1653–1670
Rowe, D.B. (1998). Correlated Bayesian factor analysis. Ph.D. Thesis, Department of Statistics, University of California, Riverside, CA
Sanchez, B.N., Budtz-Jorgensen, E., Ryan, L.M. and Hu, H. (2005). Structural equation models: a review with applications to environmental epidemiology. *Journal of the American Statistical Association* **100**, 1442–1455
Schwarz, G. (1978). Estimating the dimension of a model. *Annals of Statistics* **6**, 461–464
Song, X.Y. and Lee, S.Y. (2001). Bayesian estimation and test for factor analysis model with continuous and polytomous data in several populations. *British Journal of Mathematical & Statistical Psychology* **54**, 237–263
Thompson, B. (2004). Exploratory and Confirmatory Factor Analysis: Understanding Concepts and Applications. APA Books
West, M. (2003). Bayesian factor regression models in the "large p, small n" paradigm. In: *Bayesian Statistics*, Volume 7, J.M. Bernardo, M.J. Bayarri, J.O. Berger, A.P. Dawid, D. Heckerman, A.F.M. Smith and M. West (eds). Oxford University Press, Oxford
Zhang, N.L. and Kocka, T. (2004). Effective dimensions of hierarchical latent class models. *Journal of Artificial Intelligence Research* **21**, 1–17

Appendix: Full Conditional Distributions for the Gibbs Sampler

Suppose we have a model with k factors, conditional distributions for the PX Gibbs Sampler are presented below:

$$y_{ij} = \mathbf{z}_{ij}'\boldsymbol{\lambda}_j^* + \epsilon_{ij}, \quad \epsilon_{ij} \sim N(0, \sigma_j^2),$$

where $\mathbf{z}_{ij} = (\eta_{i1}^*, \ldots, \eta_{ik_j}^*)'$, $\boldsymbol{\lambda}_j^* = (\lambda_{j1}^*, \ldots, \lambda_{jk_j}^*)'$ denotes the free elements of row j of $\boldsymbol{\Lambda}^*$, and $k_j = \min(j, k)$ is the number of free elements. Let $\pi(\boldsymbol{\lambda}_j^*) = \mathrm{N}_{k_j}(\boldsymbol{\lambda}_{0j}^*, \boldsymbol{\Sigma}_{0\boldsymbol{\lambda}_j^*})$ denote the prior for $\boldsymbol{\lambda}_j^*$, the full conditional posterior distributions are as follows:

$$\begin{aligned}\pi\left(\boldsymbol{\lambda}_j^* \mid \boldsymbol{\eta}^*, \boldsymbol{\Psi}, \boldsymbol{\Sigma}, \mathbf{y}\right) &= \mathrm{N}_{k_j}\Bigg(\left(\boldsymbol{\Sigma}_{0\boldsymbol{\lambda}_j^*}^{-1} + \sigma_j^{-2}\mathbf{Z}_j'\mathbf{Z}_j\right)^{-1}\left(\boldsymbol{\Sigma}_{0\boldsymbol{\lambda}_j^*}^{-1}\boldsymbol{\lambda}_{0j}^* + \sigma_j^{-2}\mathbf{Z}_j'\mathbf{Y}_j\right), \\ &\qquad \left(\boldsymbol{\Sigma}_{0\boldsymbol{\lambda}_j^*}^{-1} + \sigma_j^{-2}\mathbf{Z}_j'\mathbf{Z}_j\right)^{-1}\Bigg),\end{aligned}$$

where $\mathbf{Z}_j = (\mathbf{z}_{1j}, \ldots, \mathbf{z}_{nj})'$ and $\mathbf{Y}_j = (y_{1j}, \ldots, y_{nj})'$. In addition, we have

$$\pi(\boldsymbol{\eta}_i^* \mid \boldsymbol{\Lambda}^*, \boldsymbol{\Sigma}, \boldsymbol{\Psi}, \mathbf{y}) = \mathrm{N}_k\Big(\big(\boldsymbol{\Psi}^{-1} + \boldsymbol{\Lambda}^{*\prime}\boldsymbol{\Sigma}^{-1}\boldsymbol{\Lambda}^*\big)^{-1}\boldsymbol{\Lambda}^{*\prime}\boldsymbol{\Sigma}^{-1}\mathbf{y}_i,$$
$$\big(\boldsymbol{\Psi}^{-1} + \boldsymbol{\Lambda}^{*\prime}\boldsymbol{\Sigma}^{-1}\boldsymbol{\Lambda}^*\big)^{-1}\Big),$$

$$\pi(\psi_l^{-1} \mid \boldsymbol{\eta}^*, \boldsymbol{\Lambda}^*, \boldsymbol{\Sigma}, \mathbf{y}) = \mathcal{G}\left(a_l + \frac{n}{2}, b_l + \frac{1}{2}\sum_{i=1}^{n} \eta_{il}^{*2}\right),$$

$$\pi(\sigma_j^{-2} \mid \boldsymbol{\eta}^*, \boldsymbol{\Lambda}^*, \boldsymbol{\Psi}, \mathbf{y}) = \mathcal{G}\left(c_j + \frac{n}{2}, b_j + \frac{1}{2}\sum_{i=1}^{n}(y_{ij} - \mathbf{z}_{ij}'\boldsymbol{\lambda}_j^*)^2\right),$$

where $\mathcal{G}(a_l, b_l)$ is the prior for ψ_l^{-1}, for $l = 1, \ldots, k$, and $\mathcal{G}(c_j, d_j)$ is the prior for σ_j^{-2}, for $j = 1, \ldots, p$.

Index

Lecture Notes in Statistics

For information about Volumes 1 to 137, please contact Springer-Verlag

138: Peter Hellekalek and Gerhard Larcher (Editors), Random and Quasi-Random Point Sets. xi, 352 pp., 1998.

139: Roger B. Nelsen, An Introduction to Copulas. xi, 232 pp., 1999.

140: Constantine Gatsonis, Robert E. Kass, Bradley Carlin, Alicia Carriquiry, Andrew Gelman, Isabella Verdinelli, and Mike West (Editors), Case Studies in Bayesian Statistics, Volume IV. xvi, 456 pp., 1999.

141: Peter Müller and Brani Vidakovic (Editors), Bayesian Inference in Wavelet Based Models. xiii, 394 pp., 1999.

142: György Terdik, Bilinear Stochastic Models and Related Problems of Nonlinear Time Series Analysis: A Frequency Domain Approach. xi, 258 pp., 1999.

143: Russell Barton, Graphical Methods for the Design of Experiments. x, 208 pp., 1999.

144: L. Mark Berliner, Douglas Nychka, and Timothy Hoar (Editors), Case Studies in Statistics and the Atmospheric Sciences. x, 208 pp., 2000.

145: James H. Matis and Thomas R. Kiffe, Stochastic Population Models. viii, 220 pp., 2000.

146: Wim Schoutens, Stochastic Processes and Orthogonal Polynomials. xiv, 163 pp., 2000.

147: Jürgen Franke, Wolfgang Härdle, and Gerhard Stahl, Measuring Risk in Complex Stochastic Systems. xvi, 272 pp., 2000.

148: S.E. Ahmed and Nancy Reid, Empirical Bayes and Likelihood Inference. x, 200 pp., 2000.

149: D. Bosq, Linear Processes in Function Spaces: Theory and Applications. xv, 296 pp., 2000.

150: Tadeusz Caliñski and Sanpei Kageyama, Block Designs: A Randomization Approach, Volume I: Analysis. ix, 313 pp., 2000.

151: Håkan Andersson and Tom Britton, Stochastic Epidemic Models and Their Statistical Analysis. ix, 152 pp., 2000.

152: David Ríos Insua and Fabrizio Ruggeri, Robust Bayesian Analysis. xiii, 435 pp., 2000.

153: Parimal Mukhopadhyay, Topics in Survey Sampling. x, 303 pp., 2000.

154: Regina Kaiser and Agustín Maravall, Measuring Business Cycles in Economic Time Series. vi, 190 pp., 2000.

155: Leon Willenborg and Ton de Waal, Elements of Statistical Disclosure Control. xvii, 289 pp., 2000.

156: Gordon Willmot and X. Sheldon Lin, Lundberg Approximations for Compound Distributions with Insurance Applications. xi, 272 pp., 2000.

157: Anne Boomsma, Marijtje A.J. van Duijn, and Tom A.B. Snijders (Editors), Essays on Item Response Theory. xv, 448 pp., 2000.

158: Dominique Ladiray and Benoît Quenneville, Seasonal Adjustment with the X-11 Method. xxii, 220 pp., 2001.

159: Marc Moore (Editor), Spatial Statistics: Methodological Aspects and Some Applications. xvi, 282 pp., 2001.

160: Tomasz Rychlik, Projecting Statistical Functionals. viii, 184 pp., 2001.

161: Maarten Jansen, Noise Reduction by Wavelet Thresholding. xxii, 224 pp., 2001.

162: Constantine Gatsonis, Bradley Carlin, Alicia Carriquiry, Andrew Gelman, Robert E. Kass Isabella Verdinelli, and Mike West (Editors), Case Studies in Bayesian Statistics, Volume V. xiv, 448 pp., 2001.

163: Erkki P. Liski, Nripes K. Mandal, Kirti R. Shah, and Bikas K. Sinha, Topics in Optimal Design. xii, 164 pp., 2002.

164: Peter Goos, The Optimal Design of Blocked and Split-Plot Experiments. xiv, 244 pp., 2002.

165: Karl Mosler, Multivariate Dispersion, Central Regions and Depth: The Lift Zonoid Approach. xii, 280 pp., 2002.

166: Hira L. Koul, Weighted Empirical Processes in Dynamic Nonlinear Models, Second Edition. xiii, 425 pp., 2002.

167: Constantine Gatsonis, Alicia Carriquiry, Andrew Gelman, David Higdon, Robert E. Kass, Donna Pauler, and Isabella Verdinelli (Editors), Case Studies in Bayesian Statistics, Volume VI. xiv, 376 pp., 2002.

168: Susanne Rässler, Statistical Matching: A Frequentist Theory, Practical Applications and Alternative Bayesian Approaches. xviii, 238 pp., 2002.

169: Yu. I. Ingster and Irina A. Suslina, Nonparametric Goodness-of-Fit Testing Under Gaussian Models. xiv, 453 pp., 2003.

170: Tadeusz Caliñski and Sanpei Kageyama, Block Designs: A Randomization Approach, Volume II: Design. xii, 351 pp., 2003.

171: D.D. Denison, M.H. Hansen, C.C. Holmes, B. Mallick, B. Yu (Editors), Nonlinear Estimation and Classification. x, 474 pp., 2002.

172: Sneh Gulati, William J. Padgett, Parametric and Nonparametric Inference from Record-Breaking Data. ix, 112 pp., 2002.

173: Jesper Møller (Editor), Spatial Statistics and Computational Methods. xi, 214 pp., 2002.

174: Yasuko Chikuse, Statistics on Special Manifolds. xi, 418 pp., 2002.

175: Jürgen Gross, Linear Regression. xiv, 394 pp., 2003.

176: Zehua Chen, Zhidong Bai, Bimal K. Sinha, Ranked Set Sampling: Theory and Applications. xii, 224 pp., 2003.

177: Caitlin Buck and Andrew Millard (Editors), Tools for Constructing Chronologies: Crossing Disciplinary Boundaries, xvi, 263 pp., 2004.

178: Gauri Sankar Datta and Rahul Mukerjee, Probability Matching Priors: Higher Order Asymptotics, x, 144 pp., 2004.

179: D.Y. Lin and P.J. Heagerty (Editors), Proceedings of the Second Seattle Symposium in Biostatistics: Analysis of Correlated Data, vii, 336 pp., 2004.

180: Yanhong Wu, Inference for Change-Point and Post-Change Means After a CUSUM Test, xiv, 176 pp., 2004.

181: Daniel Straumann, Estimation in Conditionally Heteroscedastic Time Series Models, x, 250 pp., 2004.

182: Lixing Zhu, Nonparametric Monte Carlo Tests and Their Applications, xi, 192 pp., 2005.

183: Michel Bilodeau, Fernand Meyer, and Michel Schmitt (Editors), Space, Structure and Randomness, xiv, 416 pp., 2005.

184: Viatcheslav B. Melas, Functional Approach to Optimal Experimental Design, vii., 352 pp., 2005.

185: Adrian Baddeley, Pablo Gregori, Jorge Mateu, Radu Stoica, and Dietrich Stoyan, (Editors), Case Studies in Spatial Point Process Modeling, xiii., 324 pp., 2005.

186: Estela Bee Dagum and Pierre A. Cholette, Benchmarking, Temporal Distribution, and Reconciliation Methods for Time Series, xiv., 410 pp., 2006.

187: Patrice Bertail, Paul Doukhan and Philippe Soulier, (Editors), Dependence in Probability and Statistics, viii., 504 pp., 2006.

188: Constance van Eeden, Restricted Parameter Space Estimation Problems, vi, 176 pp., 2006.

189: Bill Thompson, The Nature of Statistical Evidence, vi, 152 pp., 2007.

190: Jérôme Dedecker, Paul Doukhan, Gabriel Lang, José R. León, Sana Louhichi Clémentine Prieur, Weak Dependence: With Examples and Applications, xvi, 336 pp., 2007.

191: Vlad Stefan Barbu and Nikolaos Liminos, Semi-Markov Chains and Hidden Semi-Markov Models toward Applications, xii, 228 pp., 2007.

192: David B. Dunson (Editor), Random Effect and Latent Variable Model Selection, x., 170 pp., 2008.

Printed in the United States